# PICTORIAL
# GUIDE
# TO
# PLANET
# EARTH

# PICTORIAL
# GUIDE
# TO
# PLANET
# EARTH

*Frederick I. Ordway III*

THOMAS Y. CROWELL COMPANY
New York/*Established 1834*

*Designed by S. S. Drate*

Manufactured in the United States of America


**Library of Congress Cataloging in Publication Data**

Ordway, Frederick Ira, 1927-
    Pictorial guide to planet Earth.

    Bibliography: p. 177-181
    Includes index.
    1. Earth. 2. Artificial satellites. I. Title.
QB631.072                910'.02              74-34291
ISBN 0-690-62193-0

1 2 3 4 5 6 7 8 9 10

To Fred, Albert
and
Aliette Marisol,
representatives of
the first generation
to view the Earth
as it really is—
a planet.

# Acknowledgments

The author is deeply indebted to many individuals and organizations involved with Earth resources; the land, air and water environment; and overview sensing of our planet. Space prevents him from more than listing those who have been particularly helpful in making this book possible.

Robert A. Summers, Manager, Advanced Programs and Technology, Earth Observations Programs, Office of Space Science and Applications, National Aeronautics and Space Administration, Washington, D.C.; Les Gaver and Margaret S. Ware, Office of Public Affairs, National Aeronautics and Space Administration, Washington, D.C.; John Lindstrom, Manager, Nimbus/ATS Data Utilization Center and James F. Lynch, National Aeronautics and Space Administration, Goddard Space Flight Center, Greenbelt, Maryland.

James H. Sasser, Chief, Mapping Sciences Laboratory, National Aeronautics and Space Administration, Johnson Space Center, Houston, Texas; Allen H. Watkins, Manager, Earth Orbital Missions Office, National Aeronautics and Space Administration, Johnson Space Center, Houston, Texas; Edward O. Zeitler, Chief, Earth Resources Research Data Facility, Robert C. Musgrove, and R. D. Bratton, National Aeronautics and Space Administration, Johnson Space Center, Houston, Texas; John E. McLeaish, Chief, Public Information Office, Public Affairs Office, John W. King, and Bob Gordon, Johnson Space Center, National Aeronautics and Space Adminstration, Houston, Texas.

Alexander Peresich, Deputy Director, Earth Resources Laboratory, National Aeronautics and Space Administration, Mississippi Test Facility, Bay Saint Louis, Mississippi; Leland F. Belew and Joseph T. Tucker, Skylab Program Office; Charles T. N. Paludan, Environmental Applications Office, Science and Engineering, J. M. Jones, and Curtis R. Hunt, George C. Marshall Space Flight Center, National Aeronautics and Space Administration, Huntsville, Alabama; Ralph Anderson, Applications Group, and Jimmie D. Johnson, Planning and Coordination Group, National Environmental Satellite Center, Suitland, Maryland; Vincent J. Oliver, J. Gordon Vaeth, Frances C. Parmenter, James B. Jones, Donald C. Winner, and William O. West, Department of Commerce, National Oceanic and Atmospheric Administration, Washington, D.C., Rockville, Maryland and Suitland, Maryland; Robert L. Johnson, Environmental Protection Agency, Rockville, Maryland.

Charles F. Ducander, Committee on Science and Astronautics, U.S. House of Representatives, Washington, D.C.; Glen P. Wilson, Committee on Aeronautical and Space Sciences, U.S. Senate, Washington, D.C.; William A. Fischer, EROS Program Manager, Joanne K. Oman, Ruth Herbert, and Evelyn Chandler, Department of Interior, Geological Survey, Washington, D.C.; Arthur Graham, Bureau of Land Management, Department of Interior, Washington, D.C.; Bureau of Mines, U.S. Department of Interior, Washington, D.C.; Robert Williams, Bureau of Commercial Fisheries, Department of Interior, Washington, D.C.; Joseph T. Lang, Topographic Division and William R. MacDonald, Polar Operations Section, Geological Survey, Department of Interior, Washington, D.C..

Robert C. Aldrich, Principal Research Forester, Remote Sensing Unit, Department of Agriculture-

Forest Service, Berkeley, California; Mary Cowell, Photography Division, Department of Agriculture, Washington, D.C.; Audio-Visual Branch, Department of Agriculture-Forest Service, Arlington, Virginia; Robert Griffon, Jr., Bureau of Solid Waste Management, Department of Health, Education and Welfare, Rockville, Maryland; Sue F. Silverman, Special Projects Division, Federal Aviation Administration, Department of Transportation, Washington, D.C.; Franklin Pugh, Federal Highway Administration, Department of Transportation, Washington, D.C.; James Lamont, Urban Mass Transportation Administration, Department of Transportation, Washington, D.C.; Colonel Audrey E. Thomas, Department of Defense, Pentagon, OSD, Washington, D.C.; Aerospace Audio-Visual Service, Air Force Chart and Information Service, Department of the Air Force, Arlington, Virginia; Robert A. Carlisle, Still Photo Branch, Office of Information, Department of the Navy, Washington, D.C..

U.S. Naval Oceanographic Office, Suitland, Maryland; Jack Carter, U.S. Naval Photographic Center, Naval Station, Anacosta, Virginia, Hope Roberts, Office of the Oceanographer of the Navy, Alexandria, Virginia; U.S. Coast and Geodetic Survey, Washington Science Center, Rockville, Maryland; Robert Citron, Director, Center of Short-Lived Phenomena, Smithsonian Institution, Cambridge, Massachusetts; Geography and Map Division, Library of Congress, Alexandria, Virginia; James W. Moore and Mary Johnson, Audio-Visual Department, National Archives, Washington, D.C.; J. Ronald Shumate, Association of American Railroads, Washington, D.C.; Joseph F. Clayton, General Manager, Bendix Aerospace Systems Division, Ann Arbor, Michigan; Ernest E. Hardy, Center for Aerial Photographic Studies, Cornell University, Ithaca, N.Y.; Lee D. Miller, Assistant Professor of Watershed Management, Colorado State University, Fort Collins, Colorado.

I. J. Sattinger, Center for Remote Sensing Information and Analysis, Willow Run Laboratories, Institute of Science and Technology, University of Michigan, Ann Arbor, Michigan; Ronald W. Stingelin, Principal Geologist, Manager, Environmental Sciences Branch, Aerospace and Marine Systems, HRB–Singer, Inc., Science Park, State College, Pennsylvania; Terry L. Wells, Department of Development, State of Ohio, Columbus, Ohio; George E. Wukelic, Senior Physicist, Aerospace Mechanics Division, Battelle Columbus Laboratories, Columbus, Ohio.

Fernando de Mendoça, General Director, and Emmanuel Gama de Almeida, Instituto de Pesquisas Espaciais, São José dos Campos, São Paulo, Brazil; A. G. de Souza Coelho, Seçao de Fotointerpretação, Secretaria da Agricultura, Instituto Agronomicaõ, Campinas, Brazil; Luc Antoine Fobé, Projeto Sensores Remotos, Instituto de Pesquisas Espaciais, São José dos Campos, São Paulo, Brazil.

L. W. Morley, Director, Department of Energy, Mines and Resources, Canada Centre for Remote Sensing, Ottawa, Canada; A. Mayer, Natural Environment Research Council, London, England; Roberto Cassinis, Director, Laboratorio della Geofisica della Litosfera, Consiglio Nazionale delle Richerche, Milano, Italy; H. Haefner, Professor of Geography, Geographisches Institut der Universität Zürich, Switzerland; B. V. Shilin, Laboratory of Aero-Methods, Leningrad, USSR.

Should the author have neglected to have credited the assistance of any individual or organization, it is entirely unintentional. Special acknowledgment is due to Hugh Rawson for his expert editorial advice throughout the preparation of the manuscript, and to Maria Victoria Ordway for her patience in retyping a seemingly ever-evolving text. The author also thanks the many persons at NASA Headquarters; the Goddard, Marshall and Johnson centers; and the Mississippi Test Facility who provided assistance and coordination during his visits. Similar assistance was provided by the National Oceanic and Atmospheric Administration, the Department of Interior, the Department of Agriculture, the Department of Defense, the Department of Transportation, and the National Science Foundation.

*September 1974*     FREDERICK I. ORDWAY, III
*Huntsville, Alabama, USA*

# Contents

Spaceship Earth . . .

You blink your eyes and look out there;
and, you know it's three dimensional. But
it's just sitting there in the middle of no-
where. It's unbelievable.

—Astronaut Eugene A. Cernan

# PICTORIAL
# GUIDE
# TO
# PLANET
# EARTH

# 1

# The View from Above

One of the realities of living in the space age is that we must temper what we want to believe with what we reluctantly know to be the truth: that the all-important Earth on which we live is a cosmically inconsequential world lost in the immensity of space. The true size of our planet was strikingly demonstrated when astronauts first brought back from lunar orbit pictures taken at a distance of nearly a quarter of a million miles. From that incredible perspective, William A. Anders of Apollo 8 saw an

> Earth [that] looked so tiny in the heavens that there were times during the Apollo 8 mission when I had trouble finding it . . . . I think that all of us sub-

Home base as it appeared to the astronauts of Apollo 13 heading back to Planet Earth, following the explosion that nearly wrecked their spacecraft in 1970. The southwestern United States, the peninsula of Baja California, and northwestern Mexico are seen clearly. (NASA)

consciously think that the Earth is flat or at least almost infinite. Let me assure you that, rather than a massive giant, it should be thought of as [a] fragile Christmas-tree ball which we should handle with considerable care.

The new perspective that space travel has given man of his home planet more than justifies the billions of dollars that have been spent by the United States and other nations in developing rockets, satellites, and spacecraft. Not that the rocks the Apollo astronauts brought back from the Moon aren't valuable; they are. And the various scientific and communications satellites perform so well that it has taken only a few short years for technological society to come to depend on the new services that they provide. Still, in the long run it is almost certainly the new perspective and the implications that Anders —so quickly and so typically—drew from it that are of greatest significance for man's future. The lifetime of hardware, even in the vacuum of space, is measured in days and months and years, but the shift in viewpoint, though less tangible, is surely more durable, and its effects are likely to be felt for as long as men exist to sense them. The shift is comparable to the revolutionary change in the sixteenth and seventeenth centuries from the Earth-centered Ptolemaic conception of the universe to the Sun-centered Copernican view.

It is ironic today that the most ardent defenders of the environment tend to oppose national expenditures for space research when it was the first photos from space of the entire Earth that proved their case completely. Suddenly, for all men to see, there was the jewellike orb of the Earth, suspended in blackness, seemingly

so tiny, so precious, so fragile, so much— in Anders's phrase—like "a Christmas-tree ball, which we should handle with considerable care." Sometimes a picture really is worth a thousand words.

The irony of the often open, always latent split between what might be called, for lack of more precise terms, the environmentalists and the technologists, is further heightened by the fact that products of the latter offer the means— usually the cheapest, most effective means and, in many instances, the *only* means— for coming to grips with the fearful and complex problems articulated by the former.

The key is the satellite and the overview of the globe that it affords. It can monitor entire water systems, for example, tracing pollution at a river's mouth to the source far upstream. It can observe the changes that are taking place in the use of land in and around cities. It can spot forest fires— as well as areas in which trees or agricultural crops are slowly dying from disease. It can locate the most promising areas for prospecting for minerals or oil, predict the threat of floods long in advance by monitoring melting snows, or tell fishermen where to look for fish. And it can help guide ships through polar ice floes. The list of uses is practically endless. In a world that is bursting at the seams with people, where demands are increasing for limited natural resources and the environment is coming under ever greater pressures, the satellite is an essential tool for measuring, monitoring, and managing fragile planet Earth.

The advantages of artificial satellites were comprehended long before the first one was launched. However, the earliest suggestions, made prior to the era of radio telemetering, all conceived of the satellites

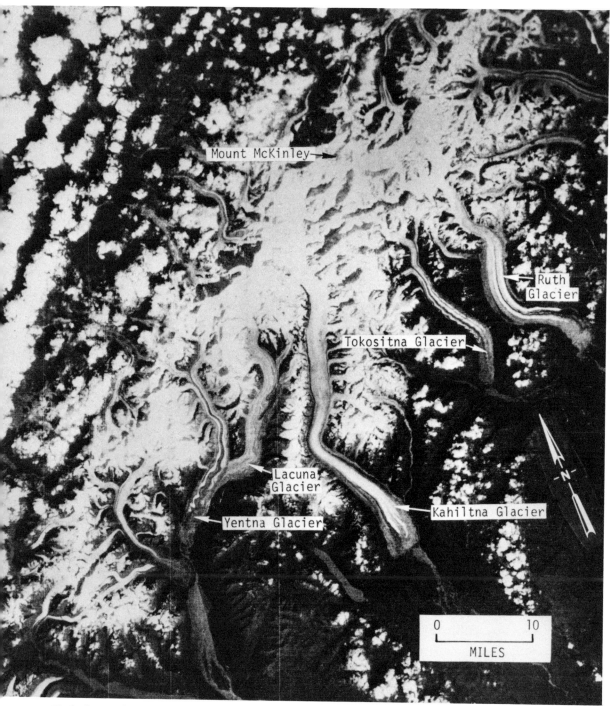

Mount McKinley→

Ruth Glacier

Tokositna Glacier

Lacuna Glacier

Kahiltna Glacier

Yentna Glacier

-N-

0          10
MILES

Tasks for satellites include monitoring glaciers, pollution, and fires. Above, "galloping" or surging glaciers, such as Yentna, Tokositna, and Lacuna in Alaska are distinguished in this ERTS 1 photo by wiggly moraines —the dark-colored strips of rock fragments they leave in their wakes. The wiggles result from alternating periods of stagnation (up to fifty years) and brief surges (one to three years) during which the ice may flow as fast as four feet per hour. Ruth and Kahiltna, normal glaciers, flow at uniform rates of a few inches or a foot per day, leaving straight moraines behind them. (NASA)

Brightness patterns on water help in assessing top soil erosion and changes in water quality, resulting from rainfall, pollution, and construction. This ERTS 1 photo shows how silt-laden waters (the lighter shades of gray) of the Delaware River (top) mix with waters of Delaware Bay (center) and empty into the ocean along the Delaware and Maryland coasts. (NASA/GODDARD)

Smoke plumes from large forest fires are readily apparent, even from an altitude of 150 miles. This photo of the northeastern coast of Queensland, Australia, was taken in 1968 during the Apollo 7 mission. (NASA)

Two historic photographs: Left, aerial view of Texas, looking south into Mexico, made from three photos taken May 24, 1954, from a Viking rocket that ascended to 158 miles, then a record altitude. Below, the first photograph to show practically the entire disk of the planet, taken from lunar orbit on August 8, 1967, by the American Lunar Orbiter 5. Distance: 214,806 miles. (U.S. NAVY AND NASA/ LANGLEY)

as manned space stations. It was assumed quite naturally that men would have to be aboard to aim and operate instruments and to observe and study the panorama below with their own eyes.

The first known suggestion for a space station came in the form of a novella by Edward Everett Hale. Appearing in 1869 and 1870 as a four-part serial in the *Atlantic Monthly*, the tale—called "The Brick Moon"—envisioned a station 200 feet in diameter placed in orbit around the Earth at a distance of 4,000 miles. Primarily constructed to serve as an aid to navigation, it was to ". . . forever revolve . . . the blessing of all seamen . . ." crossing the seas below. Hale's idea occurred almost two centuries after Sir Isaac Newton explained that, given sufficient velocity and proper aim, a body would enter into orbit around the Earth.

In Jules Verne's book *The Five Hundred Millions of the Begum*, published shortly after Hale's story (1879), Newton's concept was clearly described: "A projectile, animated with an initial speed twenty times superior to the actual speed, being ten thousand yards to the second, can never fall! This movement, combined with terrestrial attraction, destines it to revolve perpetually round our globe."

In Kurd Lasswitz's *Auf zwei Planeten* (1897) an "outside station" is introduced, set up by the Martians, one Earth radius above the North Pole. This station was not in orbit, however; it was stationary, held up by an antigravity field. Thus, neither of these cases involved a true orbiting space station; Verne had offered a satellite projectile but no station, whereas Lasswitz had proposed a station, but not an orbiting one.

The Soviet father of astronautics, Konstantin Eduardovitch Tsiolkovsky, de-

scribed the view he envisaged man would experience from a space station in his 1911 publication *The Investigation of Universal Space by Means of Reactive Devices:* "We can see how the sphere [the Earth] rotates . . . [The] picture is so majestic, attractive and infinitely varied that I wish with all my soul that you and I could see it." Later, Hermann Oberth (*Die Rakete zu den Planetenräumen*, 1923; and *Wege zur Raumschifffahrt*, 1929) suggested that orbital vehicles would be useful as communications links and refueling stations for spaceships and for observing the Earth and monitoring weather patterns. Other ideas were described in 1928 by Baron Guido von Pirquet in the journal *Die Rakete* and by Hermann Noordung (an alias for Austrian army captain Hermann Potocnik) in his book *Das Problem der Befahrung des Weltraums*. Von Pirquet proposed that three stations be established, one in a 100-minute orbit to observe the Earth, one in a 150-minute elliptical orbit intersecting both the orbit of the lower station, and that of a third, outer station to be placed in a 200-minute orbit from which interplanetary spaceships would be launched. In a much more detailed study, Noordung offered a three-element station consisting of the "Wohnrad" crew module, a power generating module and an observatory. The "Wohnrad" was shaped like a doughnut, had a 50-foot radius, and was designed to rotate around a central hub to create artificial gravity at the perimeter. The station was to be placed in a twenty-four-hour orbit which Noordung felt would facilitate observation of the Earth with the powerful optical equipment carried aboard.

Noordung's suggestions for what would now be termed an Earth resources experiment have a surprisingly modern ring.

From his observatory in space, he predicted it

. . . would be possible to perceive optical signals sent from the Earth by the simplest means, thus keeping exploring expeditions in touch with their native lands at all times. Unexplored lands could be investigated, their terrain determined, general conclusions reached about their population and their accessibility. Valuable preliminary work therefore could be done for expeditions planned, and even photographic detail maps could be furnished for new lands to be visited.

This indicates that cartography would rest on an absolutely new basis; for by means of telephotography not only could entire countries and even continents be mapped from the observatory (a task requiring otherwise many years and corresponding amounts of money), but also detail maps on any scale could be made, not surpassed in exactness even by the most scientific work of surveyors and mappers. To the latter would remain only the task of putting in contours. Above all, the still little-known regions of the earth, such as Central Africa, Tibet, Northern Siberia, the Polar regions, etc., could be mapped very exactly without much trouble.

Furthermore, important sailing routes could be kept under observation (at least by day, cloud conditions permitting), to be able to warn the ships in time about dangers such as floating icebergs, approaching storms, etc., or to announce immediately shipwrecks which had already taken place.

Since, from the observatory, the cloud movements of more than a third of the Earth can be seen at one time, while cosmic observations not possible from the Earth can be undertaken at the same time, entirely new bases for weather prediction might result.

By no means of least importance, is the strategic value of such possibilities of distant observation. Spread out like the map of a war game, there would lie before the eyes of the observer in the spatial station the entire battlefield and its approaches. Even with most careful avoiding of any movement by day the enemy would hardly succeed in hiding his plans from such "Argus eyes."

Many other proposals followed the suggestions of Tsiolkovsky, Oberth, von Pirquet, and Noordung. J. D. Bernal, in his *The World, the Flesh and the Devil* (1929), looked to the day when man would build permanent homes in space:

At first space navigators, and then scientists whose observations would be best conducted outside the earth, and then finally those who for any reason were dissatisfied with earthly conditions would come to inhabit (extraterrestrial) bases. Even with our present primitive knowledge we can plan out such a celestial station in considerable detail.

From 1930 until after World War II, few other original space-station ideas were forthcoming as attention was increasingly focused first on the development of rocket engines and later on military guided ballistic and aerodynamic missiles. Then, between 1946 and 1948, two Englishmen, H. E. Ross and R. A. Smith, developed the design of a large, rotating station to be

The glories of the land: Four remarkable views of the United States on this and the next three pages, taken during the Skylab 3 mission in 1973. (NASA)

8

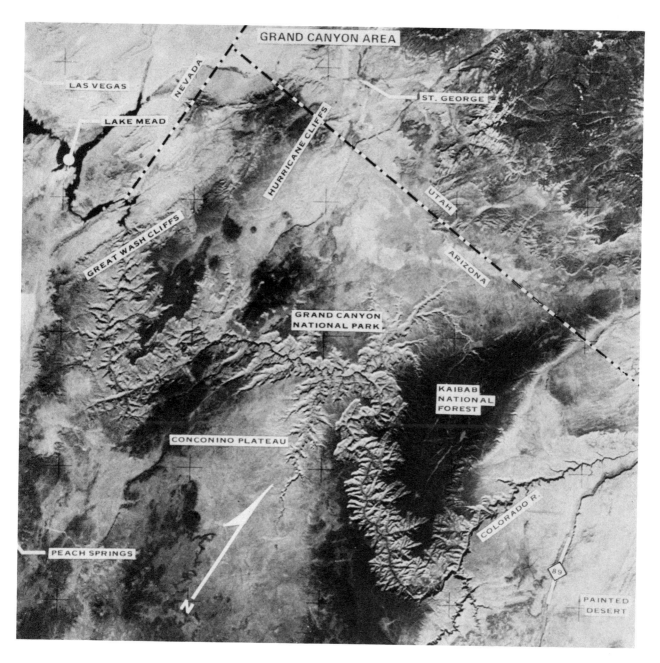

GRAND CANYON AREA

LAS VEGAS

NEVADA

LAKE MEAD

ST. GEORGE

HURRICANE CLIFFS

UTAH

GREAT WASH CLIFFS

ARIZONA

GRAND CANYON
NATIONAL PARK

KAIBAB
NATIONAL
FOREST

CONCONINO PLATEAU

COLORADO R.

PEACH SPRINGS

89

PAINTED
DESERT

N

9

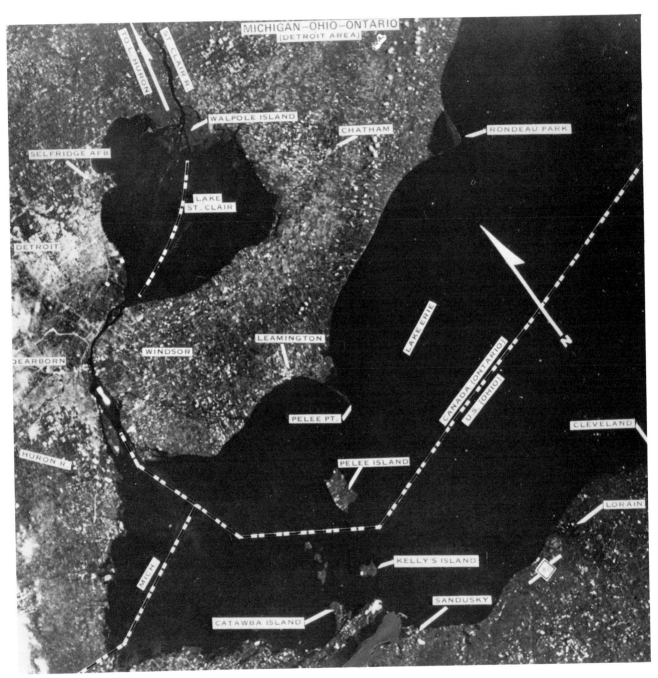

MICHIGAN—OHIO—ONTARIO
(DETROIT AREA)

TO L. HURON

ST CLAIR R.

WALPOLE ISLAND

CHATHAM

RONDEAU PARK

SELFRIDGE AFB

LAKE
ST. CLAIR

DETROIT

LAKE ERIE

LEAMINGTON

WINDSOR

CANADA (ONTARIO)
U.S. (OHIO)

DEARBORN

PELEE PT.

CLEVELAND

HURON R.

PELEE ISLAND

LORAIN

MICH

KELLY'S ISLAND

SANDUSKY

CATAWBA ISLAND

10

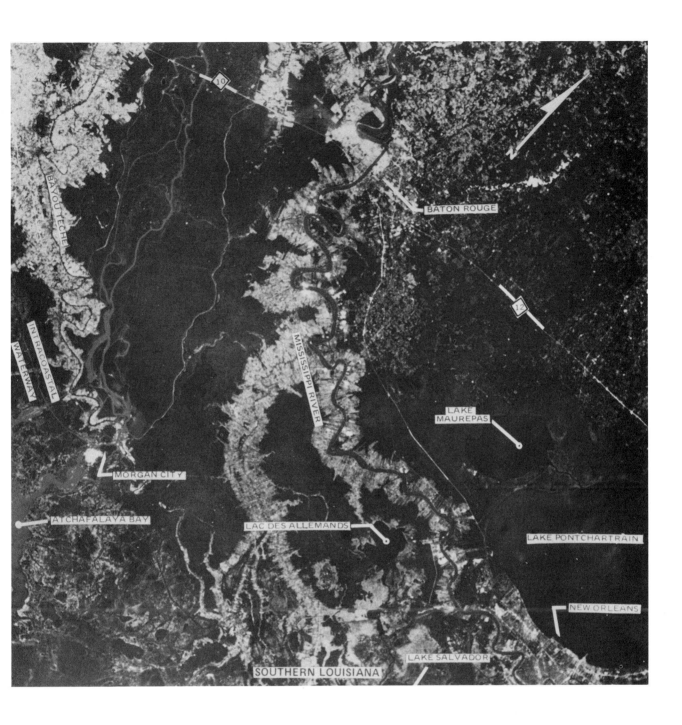

BAYOU TECHE

INTRACOASTAL WATERWAY

10

BATON ROUGE

12

MISSISSIPPI RIVER

LAKE MAUREPAS

MORGAN CITY

ATCHAFALAYA BAY

LAC DES ALLEMANDS

LAKE PONTCHARTRAIN

NEW ORLEANS

LAKE SALVADOR

SOUTHERN LOUISIANA

11

used for meteorological and astronomical research, studies of zero gravity, high vacuum physics, cosmic and solar radiation, and communications. Three years later, Wernher von Braun proposed a 200-foot, wheel-shaped station in a two-hour, 1,075-mile-high orbit primarily for use as an Earth observation platform. He wrote that

> A person observing the earth from up there would have a unique view of cloud formation on earth, particularly above the oceans. This offers novel possibilities for weather forecasting. By using high-powered telescopes, you may observe ships crossing the oceans and you may flash iceberg warnings to endangered ships. And, believe it or not, magnification factors could be used that would enable you to see people moving around on the earth's surface. This is because the atmospheric disturbances, when looking from outer space through the earth's atmosphere, are much less serious than those affecting astronomical observations from telescopes mounted on the bottom of the atmospheric shell. If we turn such a satellite telescope to the outer reaches of the universe, the planets and the stars, we shall find observation conditions which no terrestrial observatory could equal.

In the same year, Heinz Hermann Koelle presented his "Aussenstation" concept, consisting of a ring made up of 36 spheres, 16 feet in diameter, connected to a central hub by 8 supporting tubes, 4 with elevator shafts. The station would house up to sixty-five scientists and engineers and would weigh one hundred fifty metric tons. Among its many purposes would be weather observation, forecast-

ing, and control; and Earth surface observations.

Between 1951 and 1960, attention devoted to space stations increased, and in most cases so did their proposed sizes. By 1952, von Braun's station had expanded to 1,000 feet in diameter, and four years later Darrell C. Romick proposed his "Meteor," to whose 3,000-foot long and 1,000-foot diameter cylindrical terminal would be attached a wheel 1,500 feet in diameter and 40 feet thick.

Studies continued during the 1960s, but of generally smaller space stations. By 1969, NASA's attention was focused on a twelve-man integral station, based in part on a two-stage version of the Saturn carrier. This carrier offered the capability of launching a thirty-three-foot diameter payload weighing some 187,000 pounds into a 246-nautical-mile orbit at a fifty-five-degree inclination. Several integral station designs were developed, but none was realized. This approach was followed, beginning in 1970, by smaller, modular elements which could be attached to a core station or could orbit by themselves in a "free-flying" mode. It was felt that flexibility could be insured by planning for the modular growth of space stations, including the replacement of modules and the use of modules docked to the station and free-flying. Crews should be only large enough to insure a diversity of talent.

An important step in the evolution of the space station was the development and flight of NASA's Skylab. Made up of a cluster of four principal structures—a 10,000-cubic-foot Orbital Workshop (OWS), a Multiple Docking Adapter (MDA), an Airlock Module (AM) and an Apollo Telescope Mount (ATM)—Skylab was an embrionic station designed to be visited on three occasions for periods of up

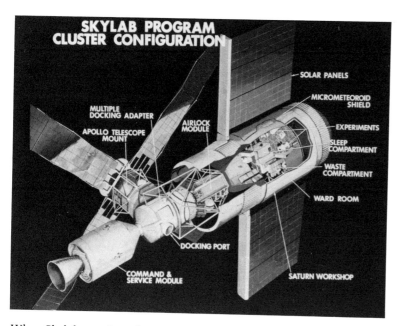

When Skylab was launched on May 14, 1973, the micrometeoroid shield and one of the two solar panels on the workshop section ripped off, so that the sight below was what greeted the first three-man crew when they ascended to the space station eleven days later. (NASA)

to eight weeks each. Once aboard, the astronauts had to conduct dozens of scientific, medical, and technological experiments—in fact, over 2,200 astronaut hours were set aside for them, which was more than three times the amount allocated to *all* previous American Earth orbit missions.

Launched by a Saturn 5 carrier on May 14, 1973, Skylab 1 was placed in a virtually perfect orbit around the Earth. (The launch itself was designated Skylab 1 and subsequent astronaut missions to the space station as Skylabs 2, 3, and 4.) Despite this success, however, the operation suffered difficulties almost from the beginning. Just 63 seconds after takeoff, the OWS's micrometeoroid shield was ripped off by aerodynamic forces. It, in turn, tore off one of the OWS solar panels and so jammed the other that it could not later deploy. The most obvious result was the immediate loss of more than half of Skylab's electric power potential. Moreover, since the shield was also designed to reflect back into space much of the solar energy that would otherwise impact on the hull, its absence caused the space station to heat up to alarming levels.

The original plan to send into orbit the astronauts the day after Skylab had achieved orbit was canceled as soon as mission controllers realized what had happened. NASA and contractor personnel worked around the clock during the next ten days to devise a thermal curtain that could be deployed by the astronauts to reduce the amount of solar energy reaching the OWS. Plans also were worked out to attempt to free the stuck OWS solar panel. After intensive training at the NASA Marshall Space Flight Center in Huntsville, Alabama, astronauts Charles Conrad, Sr., Joseph P. Kerwin,

and Paul J. Weitz were sent aloft in their Apollo ferry on May 25 by a Saturn 1B launch vehicle. Subsequently, they rendezvoused with the orbiting Skylab, made a thorough inspection of the exterior damage, and then docked. Once aboard the station, they were soon able to deploy an umbrellalike thermal protection screen, but the stuck solar panel thwarted their early attempts. Later, however, astronauts Conrad and Kerwin donned their space suits, went outside into space, and successfully released the panel by cutting the micrometeoroid shield fragment with a cable cutter tool. Within hours, the solar panel was supplying power to eight batteries inside the OWS, greatly easing the critical power situation aboard Skylab. Man the repairer had proved his worth in space.

Subsequent missions progressed flawlessly. By the time the Skylab 4 crew was recovered on the 8th of February 1974, 99 hours of Earth observation had been recorded, 60 per cent more than planned. The astronauts accumulated more than 170 mission hours and travelled 70,500,000 miles in orbit during 2,476 revolutions of our planet. The 46,146 photographs taken of the Earth below provided a dazzling source of information for geologists, geographers, meteorologists, agriculturists, oceanographers, environmental experts and other scientists all over the world.

Notwithstanding the success of Skylab and the many significant Earth observations that were carried out by preceding manned spacecraft, no doubt most of the work of orbital reconnaissance always will be carried out by the far more numerous, much less costly unmanned satellites.

Proposals for unmanned craft began to appear shortly after World War II, when

14

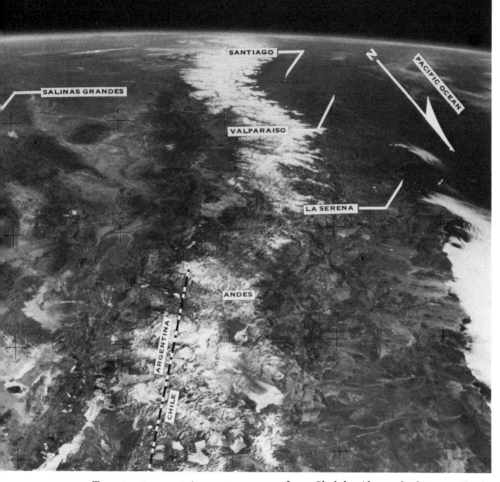

Two great mountain ranges as seen from Skylab: Above, looking south along the
backbone of the Andes; below, the intricate detail of the Alps. (NASA)

it became apparent that smaller satellites might be put to work long before large space stations could be built and orbited. K. W. Gatland, A. M. Kunesch, and A. E. Dixon proposed a minimal design in 1951, which was followed by S. F. Singer's MOUSE, or Minimum Orbital Unmanned Satellite, Earth. This 100-pound craft was to be instrumented to measure solar and cosmic radiations and magnetic fields in space; Earth observations from artificial unmanned satellites were still some years off. In order to progress in astronautics, he felt that one

> must be ready to justify a project even if the satellite is very small and minimal. Only in this way can we make use of the opportunities which the next few years may offer for such a project. If we can plan for this minimum satellite, the MOUSE, we may be launching it sooner than we now think possible.

Interest in small satellites continued, and in August 1955 Heyward E. Canney and Frederick I. Ordway, in a survey on "The Uses of Artificial Satellite Vehicles," emphasized that ". . . scientific knowledge can be given a tremendous impetus following the establishment of even a small, unmanned satellite, and that lifetimes are such as to justify the existence of close orbit minimum vehicles." These, and similar proposals in the Soviet Union, led to Russia's Sputnik 1, launched October 4, 1957, and America's Explorer 1 satellite orbited in early 1958. Since the early Sputniks, Explorers, and Vanguards, spectacular progress with instrumented unmanned satellites has permitted many observational functions to be undertaken from space with man participating on the ground.

Any satellite, large or small, simple or complex, manned or unmanned, is but one element in an often complicated information-gathering system benefiting some "user." The system normally is thought of as including the launch site; the carrier or launch vehicle; the satellite itself; tracking installations; ground receiving stations, and data links and processing centers. Occasionally, the system becomes very involved, as when a relay satellite is employed to locate and interrogate numerous information-generating platforms located in many widely separated, often remote locations. Such platforms may be balloons; airplanes; buoys at sea, on rivers, and on lakes; meteorological stations in the Arctic and Antarctic; and even migrating animals. A Tracking and Data Relay Satellite System (TDRSS) capable of providing extended coverage and service to satellites in near-Earth orbit and out to over 3000 miles high has been studied.

In order to exploit, by remote means, our world's natural and cultural resources, studies are made to determine to what extent satellites can supplement or supplant conventional observation and measuring. New or improved sensing instruments are tested (first in airplanes and later in spacecraft) and their readings checked against observations made on the ground to establish confirmatory "ground truth." Data-handling techniques are perfected to insure the best possible transmission, storage, retrieval, and evaluation. Finally, the satellite and its many subsystems evolve in the light of the requirements and limitations placed on or inherent in their application.

From its place in orbit, the satellite oversees large stretches of land and water, across many latitudes and longitudes, including such remote and long-elusive parts of the globe as the Sahara, the Antarctic, and the immense jungles of the Amazon. Its observations are synoptic, in that wide areas under the same general conditions of solar or lunar lighting are seen at once. Such observations cannot be duplicated even by high-flying airplanes. To match one satellite photo, for instance, airplanes would have to make flights over a period of months or years to obtain hundreds, perhaps thousands of photographs for compilation into a single photo-mosaic; parts of this photo-mosaic almost certainly would be taken under different lighting, weather and seasonal conditions than other parts and, more important, before the whole mosaic was completed, some parts would stand a good chance of being outdated by the changes that man and nature constantly make in the landscape.

The satellite can undertake continuous observations of the same area if it is placed in what is called a geosynchronous orbit, around the equator at an altitude of 22,300 miles; there, it will revolve around the Earth at the same rate that the Earth spins and thus remain apparently stationary over the same spot. Or, a single satellite can view parts of the entire globe in sequence over a period of time, if it is placed in a polar orbit; this will put it above a different spot on the surface after each revolution around the globe.

To avoid distortions due to differing illumination conditions, "Sun-synchronous" orbits are necessary. For a satellite to maintain the same Sun angle throughout the year, its orbital precision rate must match that of the angular rate of Earth's movement around the Sun. The altitude selected for a particular satellite is a balance between requirements for the closest view possible and wide geographic coverage, the time needed for a given observation, control of attitude and position, how and when data are being collected and relayed, and the like. Assuming that it is placed in a 500-mile high, Sun-synchronous orbit, it will be able to observe and photograph any individual target area some twenty times each year, account being taken that its period will be approximately 100 minutes during which time a target on the Earth will rotate underneath about 1,500 miles.

The satellite may observe short-term, even one-time phenomena—a volcanic eruption, a flood, a storm—as well as long-term and seasonal changes—the meanderings of a river bed, the formation of shoals before a harbor, the progression of snow lines. The observations of the satellite are not limited to the visible spectrum, but depending on the instruments they carry, may be conducted throughout the entire electromagnetic spectrum, from the ultraviolet, through the visible, the infrared, passive microwave, radar, and X-ray wavelengths.

The various sensors aboard unmanned satellites are, in reality, signature detectors. Every object has its own electromagnetic signature—that is, it emits or reflects waves in a distinctive way that serves to identify it as if its characteristic wave pattern was a fingerprint. A signature reveals not only the nature of an object but its condition. Thus, while the signature of water differs from that of rock, the signature of warm water or choppy water differs from cold water or calm water. And in the plant world, each agricultural crop has its own distinct "slot" or characteristic signature in the electromagnetic spectrum. The slots can vary, depending upon whether the plants are healthy or not. For example, observations in the near-infrared

17

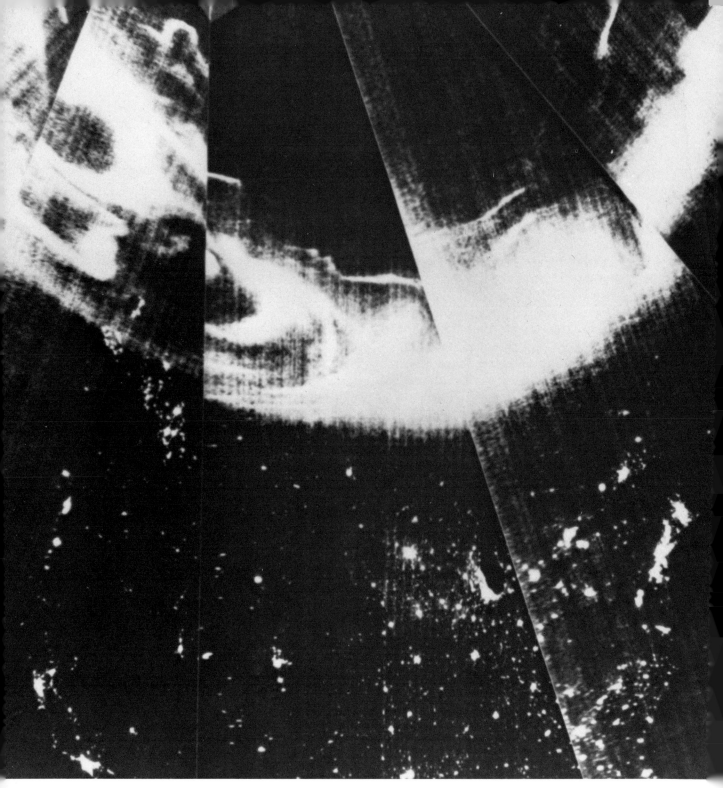

Two remarkable nighttime views of the continental United States, both made by U.S. Air Force satellites, from about 500 miles altitude. In the composite photo above, with essentially no reflected moonlight, the entire country is limned by the light of its cities, while to the north, the Aurora Borealis shines spectacularly. The photo at right, taken on a cloudy but moonlit night, shows the eastern half of the country, with the lights of the major cities. The whiteness around the eastern Great Lakes is snow on the ground. (USAF)

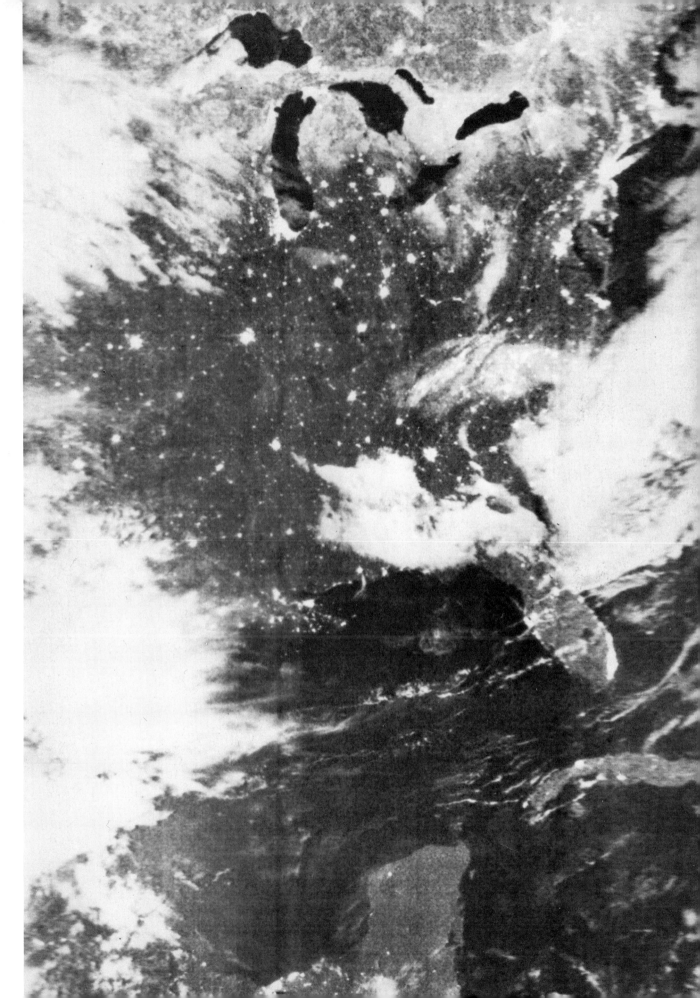

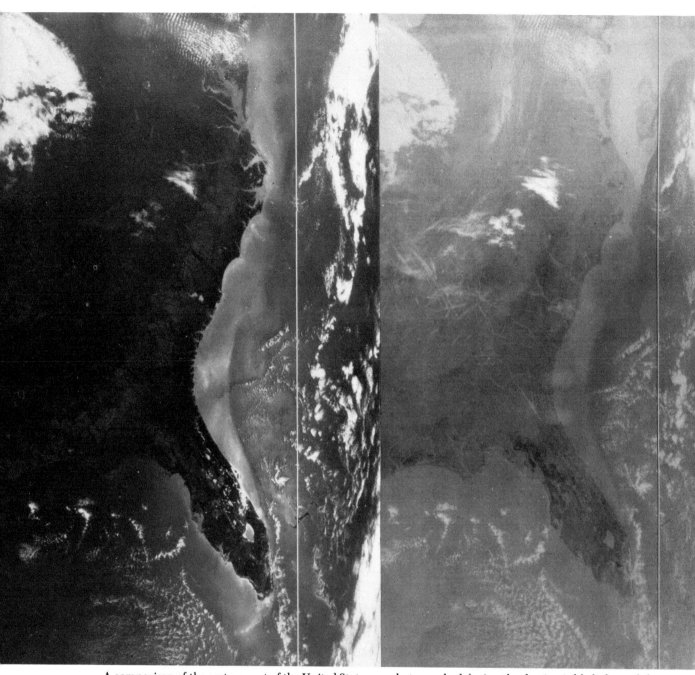

A comparison of the eastern part of the United States, as photographed during the day in visible light, at left, and during the night by infrared, right. Both images were made in April 1973 by the NOAA weather satellite. (NOAA/NESS)

20

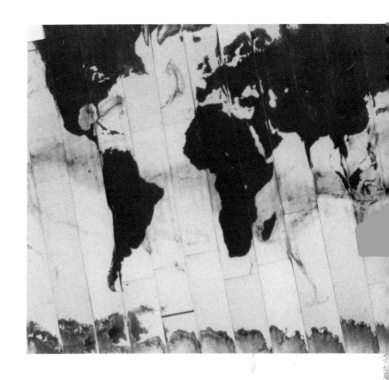

A portrait of the globe made by microwave, and so unhampered by clouds, this montage was produced on thirteen consecutive orbits by Nimbus 5 in January 1973. The land areas appear darker than the oceans because they reflect more microwaves. In the polar areas, new ice and snow appear darker than old ice and snow; over the oceans, areas of intense rain appear dark while intermediate shades of gray indicate moderate rainfall. (NASA)

can lead to the detection of loss of vigor in wheat fields plagued with "wheat rust" (*Puccinia graminis*) or suffering from inadequate fertilization or irrigation. In order to image the maximum tonal contrast signature between the vigorous and unhealthy wheat, a near-infrared-sensitive film with a deep red filter is generally employed—the latter to prevent undesirable wavelengths from registering on the film emulsion. Developed photos show that a vigorous wheat field is lighter than those that are diseased. On some plants, an unhealthy condition leads to a loss of chlorophyll, which is also measurable by this means.

Agriculturists and forestry users of satellites concentrate in the near-infrared region, while oceanographers employ somewhat longer wavelengths, associated with the thermal radiation from the ocean. At still longer wavelengths, in the microwave spectrum, passive radiometers provide such information as the temperature of the soil, while active microwave side-looking radar systems are excellent for all-weather coverage of topography and geologic features.

Two types of infrared sensors are used: radiometers that measure radiations from the surface below yielding a graphic reading and scanners that scan in a perpendicular fashion giving rise to surface images. Of rapidly rising importance is multispectral (or multiband) photography, wherein several black-and-white pictures are simultaneously exposed through narrow-band filters, later to be combined in the laboratory to form what is known as composite false-color photographs. Filter bands and printing techniques are selected that give the highest color contrast between the various elements under investigation. Taking agricultural applications as an example, three spectral bands are generally used: (1) the blue green part of the spectrum to permit haze penetration, shallow water penetration to map underwater vegetative growths, and land form mapping; (2) the red and near-infrared parts to reveal vegetation vigor and moisture distribution; and (3) an intermediate band to provide for crop recognition. Each and every type of growth must be related or considered in terms of the season; time of day; illumination; weather conditions between sensor and surface; quality of soil nutrients; local moisture conditions; disease, insect and other stresses, and state of growth (health). By examining visual images in combination with images produced by sensors using several frequencies, tone and texture differences become evident and many types of plant life can readily be identified. Multispectral scanning distinguishes the diseased plant from the vigorous one because of differences in reflected or emitted radiation. Some infrared sensors can help evaluate crop response to different fertilizers used in varying quantities, as well as to pesticides, fungicides, and insecticides.

NASA's Earth Resources Survey Program has three principal elements: (1) the Earth Observations Aircraft Program, (2) the Earth Resources Technology Satellite (ERTS) program, and (3) the Earth Resources Experiment Package (EREP) carried aboard Skylab in 1973 and 1974. Each of these data-gathering programs is backed up by field and laboratory activities conducted not only by NASA but by other interested agencies and their university and industrial contractors. In addition, studies are being made toward the definition of Earth observation experiments to be tied to future space shuttle, Spacelab and space station programs.

The aircraft program got started in late 1964 when a Convair 240A was outfitted as a test bed for sensor development. A few years later, a Lockheed NP-3A Electra was introduced followed by a NC-130B Hercules aircraft. Then, NASA started flying an RB-57F Canberra jet that allowed sensors to operate from above 90 percent of the Earth's atmosphere. Until it crashed as a result of a midair collision in April 1973, the space agency also employed on occasion a Convair 990 research aircraft named *Galileo* based at NASA's Ames Research Center. Its most publicized Earth resources assignment was in connection with the cooperative American–Soviet survey of the Bering Sea during the winter months of 1973. It was later replaced by Galileo II which, beginning in mid-June 1974, joined 12 other aircraft, 38 ships, 60 ocean bouys, 6 types of artificial satellites, and almost 1,000 land stations in an international program to gather data from the top of the atmosphere to sea level and down to about a mile below the surface of the ocean. Called GATE, for GARP (Global Atmospheric Research Program) *Atlantic Tropical Experiment*, it took place over about 20 million square miles spanning the Eastern Pacific, South America, the Atlantic Ocean, Africa, and the western part of the Pacific Ocean. Supporting the six-month operation was NASA's most advanced weather satellite, the Synchronous Meteorological Satellite 1, which had been placed into a 21,471-mile orbit over the equator only a month before GATE began.

Commercial enterprises have also entered the Earth resource arena, notably Grumman Ecosystems, the Aero Service Corporation (part of Litton Industries),

and Goodyear Aerospace. Using Gulf-stream 1 and A-260 aircraft, Grumman has conducted (for Petroleos del Peru) side-looking radar and magnetometer sensing of some 135,000 square miles of Peruvian jungle in connection with an oil exploration program; and, under contract with the United States Geological Survey, side-looking radar coverage of the states of Massachusetts, Connecticut, and Rhode Island. The most impressive feat of Aero Services and Goodyear Aerospace was the overview mapping of large tracts of Brazil and Venezuela using a modified Caravelle jet equipped with side-looking radar and other devices. Since much of the area is

The Boston area, from an airplane and from Skylab. Above, the view from 51,200 feet, made by an aircraft of the Earth Observations Division of NASA's Manned Spacecraft Center. Below, the same general area, from almost 270 miles, as seen from Skylab in 1973. (NASA)

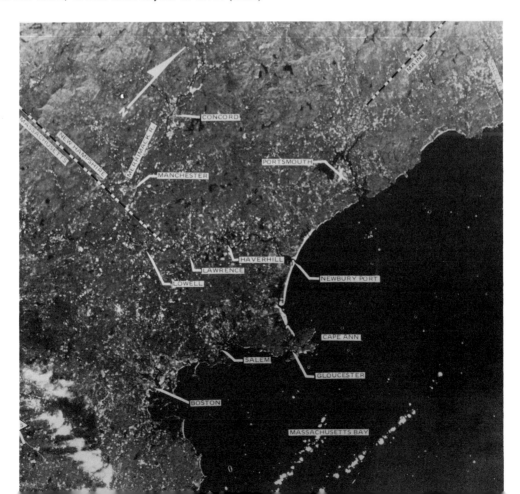

covered by clouds for long periods of time, the cloud-penetrating side-looking radar was found to be very effective for geological and other purposes.

From 1972 through 1974, virtually all EOAP aircraft were assigned supporting roles in the ERTS 1 satellite and Skylab programs; for example, during 1972 alone approximately 150 overflights were made of ERTS 1 test sites. Aircraft also played key roles in disaster surveys in cooperation with the National Oceanic and Atmospheric Administration and the State of California. Among them: surveys of two forest fires, flooding resulting from a broken dike, and floods induced by hurricane Agnes.

Selected "multidisciplinary" sites are overflown repetitively in the United States as remote sensing studies are made in the fields of geography, geology, hydrology, agriculture, forestry, oceanography, and marine biology. Other sites are examined strictly from ecological and environmental points of view; for example, NASA aircraft are undertaking wildlife management studies of the entire state of Arizona and are helping its Department of Transportation to develop a bicycle and footpath network. And, in cooperation with ground stations and radiosonde balloons, other NASA aircraft are measuring the composition and tracing the movements of smog over the San Francisco Bay area and the Livermore and Santa Clara valleys. NASA planes have overflown sites in Brazil and Mexico; and, in support of the International Biological Program, flights were made over Argentina's grassland. At the request of the Peruvian government, photographs were made of the results of the great May 31, 1970, earthquake that so devastated that South American country.

The first U.S. spacecraft devoted exclusively to Earth resources is named, appropriately, Earth Resources Technology Satellite No. 1, or ERTS 1. A firm program to orbit two such craft was approved in January 1969; and, in July 1970, NASA selected General Electric to develop and construct them. ERTS craft were designed to transmit all data to ground receiving stations by telemetry; no provision having been made for physical recovery of film. ERTS 1's orbit was near-polar, Sun-synchronous, 103 minutes long, and between 560 and 565 miles high. This altitude gave a ground swath— from the point of view of the instruments—of 100 miles on a side, meaning that each ERTS 1 image covered 10,000 square miles of land or sea. Accuracy was such that features less than 300 feet in size could be detected readily. The ground coverage interval varied from 18 days at a typical site on the equator to 9 days at 45 degrees north and south latitudes, to even less further poleward.

ERTS 1 was orbited on July 23, 1972, and almost immediately began to accumulate data. Among its various instruments were a high-resolution multispectral television system known as the *return beam vidicon;* a multispectral scanner; wide-band videotape recorders; and a data collection system. By taking pictures in four different color bands and combining them in varying ways, it was possible to obtain more knowledge about a particular scene or condition on the surface than if a common color photograph had been taken of it. Moreover, this "false color" technique was readily compatible to computer enhancement and other techniques indispensable to the Earth resources analysis and evaluation program. All information gathered by the satellite was stored,

The heart of France, in an ERTS 1 photo, taken November 18, 1972. Paris is easily visible in the upper left; the Seine loops across to the east, where it is obscured by clouds. At the lower right is the Loire. (NASA)

England, Wales, the English Channel, and, in the distance, the North Sea, photographed during the Skylab 2 mission. (NASA)

subsequently transmitted to receiving stations in the continental United States and in Alaska. Instructions to the satellite, meanwhile, were sent from an Operations Control Center. Processed data were made available to users through the NASA Goddard Space Flight Center in Greenbelt, Maryland. Typically, information obtained was employed to produce photo images at the 1-to-1,000,000 scale, of great interest to agriculturists watching seasonal changes in vegetation, to geographers, and most particularly to geologists working on studies involving regional-scale structures. In order to coordinate the immense mass of information flowing from ERTS 1 (and also from Skylab), a Federal Earth Resources Survey Program operated by an Interagency Coordinating

Committee was established early in 1972. Members were chosen from NASA; from the departments of Agriculture, Commerce, Interior, Defense, and State; from the Environmental Protection Agency; and from the U.S. Army Corps of Engineers.

More than 300 experiments were conducted with ERTS 1, many of them carried out by non-U.S. investigators from 37 countries and two United Nations agencies. Of the disciplines involved, geology accounted for the largest number (68), followed by multidisciplinary resources (46), agriculture-forestry range resources (42), water resources (41), and land use survey and mapping (32). Twenty-eight investigations were applied to the environment, 19 to the ocean and the rest to interpretation techniques development, sensor technology, and meteorology.

Seven and a half months after orbiting, ERTS 1 had covered some 90 percent of the United States, cloud-free, and about 75 percent of the Earth as a whole. Charles W. Mathews, NASA's associate administrator for applications, reported in March 1973 that ERTS 1 had produced positive results in nearly every area that had been proposed for it:

> Specifically, we can say with assurance that many of the major crop species can be identified well enough for inventory from space: that forest fire and flood damage, even in the remotest areas, can be economically and quickly assessed; that snow surveys can be made with sufficient precision to aid in the control of hydraulic power from dams; that the data can be automatically transformed into usable map products including land use maps, and that new geological features can be found even in well-mapped areas.

Although Earth observation experiments were undertaken in the U.S., Mercury (1963), Gemini (1965–1966), and Apollo (1967–1972) spacecraft programs and comparable Soviet Earth-orbiting craft, it was only after the

ERTS photographs have very little distortion because they are taken from so high up—about 565 miles—with the result that they can be fitted together into nearly true flat cartographic maps. This mosaic of the U.S. East Coast, from Cape Cod to the Florida Keys, was made from parts of thirty-eight color pictures, each of which in turn was built up by photographing three black-and-white images through color filters. (NASA)

launching of the Skylab space station on May 14, 1973, that man became able to direct and monitor such observations for periods measured in weeks instead of days. The orbiting laboratory incorporated an Earth Resources Experiment Package, or EREP, which was designed to take maximum advantage of experience accumulated by earlier manned satellites as well as by such unmanned craft as ERTS 1. Incorporating five experiments, EREP permitted simultaneous sensing of ground truth sites in the infrared and microwave regions of the spectrum. Data received were compared with information being collected at the same time from unmanned spacecraft, from aircraft, and by ground observers. This permitted signatures sensed from Skylab to be validated. The astronauts on board were occupied first with the acquisition of both primary and secondary targets and subsequently with the operation of equipment and with seeing that the collection of non-wanted data was minimal. They also changed experimental procedures and sequences when it was deemed necessary.

Skylab's multispectral photography facility consisted of six 70-mm. cameras all aimed at the same target at the same time, each frame taken covering approximately 80-by-80 nautical miles with a resolution of about 100 feet. Also carried was an infrared spectrometer (called a filter wedge spectrometer) capable of sensing surface phenomena in the visible, near-infrared, and far-infrared spectral regions; a multispectral scanner experiment; a microwave radiometer/scatterometer altimeter, and an L-band microwave radiometer designed to measure the surface brightness temperature.

Manned Earth resource surveys are scheduled to continue during the post-Skylab era dominated by the space shuttle. Shuttle payloads called Spacelabs are to be employed first, perhaps later to be followed by modularized space stations. The Spacelab, which is under joint development by the United States and the European Space Research Organization, is a shuttle-launched, pressurized module to which is attached an unpressurized platform ("pallet") designed to accommodate large instruments and sensors. The main laboratory will house experiment control stations, smaller instruments, data recording systems, and various subsystems that augment the principal systems in the shuttle. Earth resource and other studies will be conducted from the European Spacelab nestled in the bay of the shuttle's second (orbiter) stage. For experiments requiring orbits higher than those achievable by the shuttle-lab combination, a space tug will be employed. When developed it should be capable of deploying, servicing, and later recovering detached experiment modules—which could be manned or unmanned.

Seven areas of experimentation have been identified for post–Skylab manned Earth resource activities. These include (1) meteorology and atmospheric sciences, (2) world land-use mapping, (3) air and water pollution, (4) resource recognition and identification, (5) natural disaster detection and assessment, (6) ocean resources, and (7) special research (for example, the study of the use of laser as a scanner, the direct transmission of high-resolution television imagery, and certain experiments that involve the concurrent use of ground, airplane, sounding rocket, and weather satellite-mounted sensors). In all seven areas, astronauts will be trained to install, check out, calibrate, and align sensors; operate them and observe their performance; make "quick-look" evaluations of the data being produced; shut down and store equipment when necessary; be responsible for data discrimination and transmittal; perform maintenance, repair and replace parts as necessary; continuously consult, via radio link, scientists and others responsible for individual experiments; and help plan continuing mission objectives.

A particularly intriguing objective of future Earth resources studies will be to watch for the occurrence of events that are not only unpredictable but of very short duration. At the present, such events are generally reported from ground and air observers to the Center for Short-Lived Phenomena established by the Smithsonian Institution in Cambridge, Massachusetts, in January 1968. Short-lived phenomena range from millions of blackbirds flocking at one time into a very small area, to huge concentrations of house mice that once reached an average density of 200 per acre in the Australian state of Victoria; from the sudden appearance of volcanoes and even volcanic islands to the just as sudden disappearance of glacial lakes; from ecologically disastrous oil spills to the clandestine dumping of chemicals into rivers; from severe storm erosion to earthquakes; from natural-gas eruptions to landslides. The oil spill problem may well become a major observational activity of orbiting astronauts in the years to come, as super-tankers like the three-million-barrel capacity *Nisseki Maru* (which requires 89 feet of water to float) ply the world's oceans.

Almost exactly two centuries will have passed from the time man first took to the air in balloons to the space shuttles, the space laboratories, and the space stations of the 1980s. Approximately one century will have elapsed following the publication of the stories of Hale, Verne, and Lasswitz, and about fifty years from the time the massive German missile program got under way in the early 1930s that led to the V-2. Not all the early space enthusiasts recognized that one of the prime beneficiaries of astronautics would be the very Earth they were so intent on leaving. But, as Professor A. M. Low remarked nearly forty years ago, "If anyone still believes the use of rockets or the facts of space travel to be foolish, let him remember that to neglect those possibilities, however vague, would be an even greater folly."

Virtually all of southern Greece, the Gulf of Corinth, and about half the northern peninsula as recorded in a single photograph during the Skylab 3 mission in 1973. (NASA)

# 2
# Geography— Twentieth Century Style

The land is spectacular in its diversity, with its huge mountains, remote deserts, and vast plains; its inland waters, forests, and megalopolises. Surprisingly, in this the twentieth century, when the age of exploration is long since past, many parts of this marvelous globe are poorly and inaccurately mapped. No continents or islands remain to be discovered, but much filling in must be done before our understanding of the land surface can approach completeness.

The question of maps is by no means academic. The exploding world population and the rising expectations of nearly all individuals in the face of dwindling proven reserves of natural resources, combine to make the creation and maintenance of truly modern maps one of the most important tasks of this decade.

Geographers estimate that somewhat less than one-half of the world's land surface has been adequately mapped; some areas have not been properly mapped at all, others have received only minimal attention. Not surprisingly, the developing countries tend to have the poorest maps—along with the greatest needs for the natural resources whose discovery or exploitation has been prevented or inhibited by lack of the basic maps.

Developed countries, too, often lack adequate maps. Some parts of the United States have not been aerially photographed in twenty years and many so-called modern maps are a decade old, which makes them dangerously obsolete, given the rate at which the face of the land is changed by natural forces and especially by the works of man.

Map-making is only one of the concerns of the geographer. In broad terms, he is responsible for the study of the whole of the planet's surface, from small Pacific atolls to the seventeen-million-square-mile Asian continent. He studies not only the arrangement of physical features but the interrelationships of shapes and processes and the correlation of plant and animal life with the varied and constantly changing surface. Because of the immense reach of these studies, geographers tend to specialize in such subfields as physical geography, economic geography, demography or population geography, settlement and urban geography, and cartography or map-making.

Geographers realized what a boon the satellite overview was to their work and began taking advantage of it soon after the space age opened with the launching of Sputnik in 1957. Cameras were sent aloft in manned spacecraft in the American Mercury and Gemini programs. The Soviets made comparable experiments. Some unmanned spacecraft, meanwhile, carried automatic cameras; pictures taken this way were delivered to the ground in a

reentry capsule jettisoned from the spacecraft. The reentry capsule would be either "snatched" by an airplane as it descended through the atmosphere by parachute (now a routine, highly reliable method of retrieval in the U.S. Air Force) or recovered after landing.

More common than the use of reentry capsules today, however, is the transmission of pictures and other geographical data to the ground by radio or television. Pictures may show anything—from isolated hills and mesas to whole mountain chains, from individual orchards to huge jungles, from a single snow field to a vast continental glacier, from isolated hamlets to sprawling urban agglomerations. These, and innumerable other geographical features, are readily identifiable by a variety of characteristics that can be sensed from above and transmitted to ground control stations on command or at preset times.

Many land phenomena, it turns out, can be examined in more than one spectral region, leading to composite images useful to geographers. For example, thermal sensing in the infrared region is applicable to such short-lived and relatively uncommon events as volcanic eruptions; microwave sensing can help to make all-weather and nighttime imaging practical, and spectral-signature techniques are sensitive to such factors as diurnal, climatic, and seasonal variations in energy absorption, emissivity, and reflectivity.

The satellite overview is leading to great improvements in map-making. Coverage from conventional aerial mapping is almost always discontinuous and the altitude limitations of the aircraft prevent synoptic views of the land below. In areas of difficult access, coverage is usually

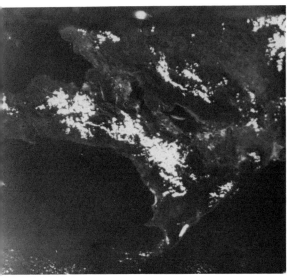 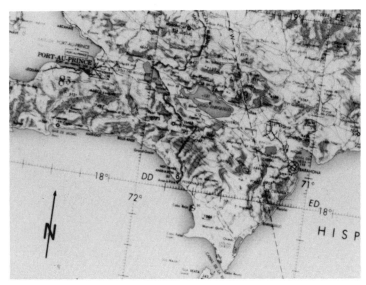

Improving maps through space photography: Many details of lake and ocean shorelines, stream channels, and other topographic features were rapidly and inexpensively transferred from the Gemini 7 photo above at left of parts of Haiti and the Dominican Republic to the map at right of the Port au Prince area. The photograph was taken December 13, 1965 from an altitude of about 150 miles. Below, the first map of the entire continental U.S., a mosaic of 595 ERTS 1 images, taken from July 25 to October 31, 1972. (NASA)

The Straits of Gibraltar, once thought to be the end of the world, as seen from Apollo 9 in 1969. (NASA)

spread across not only seasonal boundaries but time intervals that may approach decades. To construct a small-scale map of the United States from aerial photographs would take at least several years since five hundred thousand to one million photographs would be required for the photo mosaic.

A satellite in orbit, in contrast, can do the same job in a matter of weeks with four hundred to five hundred photographs. ERTS 1, for example, took thousands of cloud-free photos each covering an area of more than ten thousand square miles.

Geographers (and other scientists utilizing geographic data) require timely information for many reasons. Planners need to know the actual boundaries of cities and towns, to distinguish among the urban areas devoted to residential, industrial, and commercial pursuits, and to assess continually the encroachment of all three into the countryside. According to Arch C. Gerlach, chief geographer of the U.S. Geological Survey, "By using different cameras trained on urban and surrounding areas, it will be possible to identify changes in geographic features, transportation linkages, urban growth and functional changes, and other characteristics and trends difficult to observe with conventional photos."

At the Fourth Annual Earth Resources Program Review held in January 1973 at NASA's Manned Spacecraft (now Johnson Space) Center at Houston, Gerlach predicted that within four years it would be possible through ERTS 1 photography to delineate land cover and land use for the entire country at a scale of 1:250,000, which could then be used "for national and regional planning, for monitoring and measuring trends, and hopefully for

A black-and-white version of a color infrared photograph portrays land uses near Wilmette, Illinois: (A) modern, single-family-type housing, (B) new apartment buildings, (C) golf course, (D) farm land being withheld from development, (E) cemetery. (NASA/NORTHWESTERN UNIVERSITY)

predicting and evaluating desirable modes of future land use." As an interim step, the Soil Conservation Cartographic Division of the Department of Agriculture released (in April 1974) a giant photo mosaic of the entire continental United States, the first ever made from satellite imagery. Measuring 16 by 10 feet, it was produced from 595 cloud-free images taken by ERTS 1 at an altitude of 560 miles under the same lighting angle. The map, referred to as a controlled-base mosaic since the images had to be corrected for accuracy and scale, was produced at a scale of 1:1,000,000.

35

Of course, repetitive overview coverage will show rates, directions, and time sequences of decay as well as of growth. In New York City alone, it is estimated that 20,000 dwellings are abandoned each year, and in virtually all U.S. central cities vacant areas appear with increasing frequencies as buildings are demolished and not replaced. By having access to inputs from all corners of the globe, a composite picture of the world movement, of population, and of the effects they have on the land can be prepared. The practical value of ERTS 1 photographs to urban planners was described in some detail in a statement before the United States Senate Committee on Aeronautical and Space Sciences in March 1973 by Dartmouth College geographer Robert B. Simpson. He and his co-workers discovered that they "could consistently distinguish between multi-family, single-family and mixed-housing, and could spot communities as small 700 persons" in the southern New England area. They also could identify industrial and commercial tracts as small as 800 square feet in area as well as major highways and many smaller roads passing through woods, golf courses, bridges, and small canals. Noting that his group is using ERTS 1 imagery to focus on "the problems of urban development and the spread of cities," Simpson declared that "If we could have had ERTS imagery 25–30 years ago, we probably would have much more green space between Washington, D.C. and the Boston area. . . . The zoning practices around the cities along the Eastern seaboard probably would have been a lot different." His data are used for planning in the Boston and New Haven regions in accordance with a four-stage sequence: "The first stage will be to determine what the urban landscape is.

The next step will be to determine why this imagery looks the way it does. The third step will be to predict future trends. The final step will be to take the necessary action required for the best use of the land."

In Arizona, meanwhile, space photographs have been used to update aerial photos of the Salt River Indian Reservation. According to the Bureau of Indian Affairs, the overview led to the discovery that several economically valuable sand and gravel pits were within—rather than outside—the reservation boundaries. Moreover, two stream diversion programs were halted when it was determined that the changed courses would flow over Indian land.

In Brazil, space imagery of the Amazon basin produced "extraordinary" results, in the opinion of the general director of Brazil's National Space Institute, Fernando de Mendonça:

The course of the tributaries of the Amazon river are very different from the ones shown in the most recent available charts. . . . Islands with more than 200 km² area exist which are not shown on maps. . . .Small villages and towns are located wrongly on the maps by several tens of kilometers.

In particular, de Mendonça noted errors in road building:

The drainage systems of some areas are entirely wrong and this has caused, among other things, the construction of roads (Manaus-Porto Velho, for instance) with extra expenditures for bridges. In fact, the mentioned road is placed wrongly (by more than 20 km) in recent maps (1971) . . .

36

Washington, below, to Baltimore, top, viewed from Skylab 3 in 1973. The Chesapeake Bay is to the right (east), the Potomac winds through the capital in the lower left (southwest). The extent of urbanization, major highways and bridges, and other features, including harbor and tunnel facilities in Baltimore, the mall area in Washington, and the Capitol Building can be discerned. On the land, cleared areas appear lighter than wooded areas; in the Bay, contrasting shades show circulation patterns, with darker areas indicating clearer waters. (NASA)

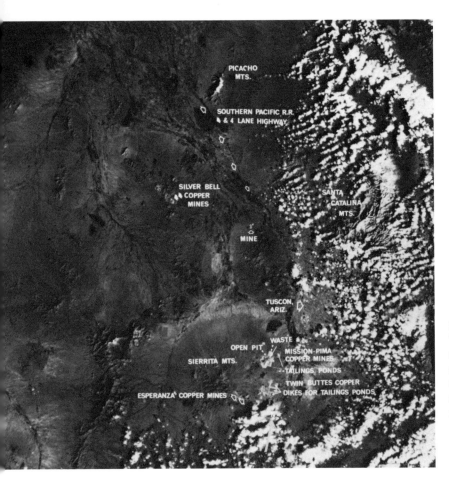

Mine dumps and mill tailings ponds near Tucson, Arizona, can be easily identified and their cultural effect assessed through space photography. This Apollo 9 photo shows an area approximately 80 miles square. (U.S. DEPT. OF INTERIOR/GEOLOGICAL SURVEY)

Roads such as this frequently are built without adequate preparation and with little or no consideration for their long-range impact on the environment and the human condition. An excellent example is another Brazilian enterprise, a highway thrusting straight through the heart of the great Amazon rain forest. Running approximately parallel to and about 200 miles south of the river, the Transamazonian Highway was started with only hasty, pre-ERTS 1–Skylab preparatory surveys and apparently with scant concern for its effects on the primitive Indian population, on the region's incredibly diverse wildlife, on the terrain that eventually will be exposed by the settlers that follow the bulldozers, and on the climate of the entire Amazon basin—perhaps on the entire world.

When completed, the highway will be about 9,000 miles long and will represent an investment upward of $500 million. An overview photograph of the 2,700,000-square-mile Amazon basin crossed by the Transamazonian Highway graphically shows its potential importance as "the dorsal spine of Brazil"—in the words of Minister of Transport Mario Andreazza. As the highway punched westward toward the Peruvian border, hordes of almost impoverished Brazilians followed, lured by the government's offer of 250 acres gratis, a modest house, and family loans of more than $2,000. As the settlers moved in, the results began to become evident on orbital photographs. Spur roads branched out from the main highway. Buildings proliferated and, most ominously, trees began to disappear.

The planetary consequences of the destruction of the vast Amazon rain forest are of great concern not only to those with strong esthetic persuasions but to environmentalists as well. Planners responsible for the Transamazonian Highway and the regional development it inevitably will create estimate that by the year 2050 the whole Amazon jungle will have disappeared. Sensors aboard orbiting satellites will watch this inexorable process, which may bring about important changes in climate not only in Brazil but all over the world. Moreover, as the jungle goes, the thin topsoil will be exposed to the elements and its nutrients will rapidly disappear. A veritable desert may result, adding to rather than alleviating regional poverty. Perhaps the orbital monitoring of these changes will bring about corrective actions before the point of no return is reached, simply by providing regular and timely information on what is happening below. If climatologists, for example, demonstrate with reasonable certainty that the world climate will be detrimentally affected by the disappearance of the Amazon rain forest, international pressure may be brought to bear to attempt a halt to the development. Brazilian activities in the Amazon prompted Bengt G. Lundholm of the Swedish Natural Science Research Council in Stockholm to remark: ". . . to an ecologist the exploitation of the Amazon Basin without any knowledge about its ecological consequences can be a global disaster." This is but one example of how the actions of man on geography and the environment of one part of the world may adversely affect countries thousands of miles away.

Global concern over the practical as well as the aesthetic implications of man's urge to despoil nature rather than to live within its bounds is certain to increase—as will reliance on the geographical overview. For example, everyone is familiar with the visible pollution produced by gasoline-powered cars and by factories powered by coal, but when we turn on the lights or put on the electric oven there is no visible pollution, which is principally created at a distant generating plant. But the satellite can remind us what is happening.

An excellent case in point is the great southwest power consortium that aims to create what could become the second largest power complex in the U.S., producing some 14 million kilowatts of electricity (nearly the capacity of the Tennessee Valley Authority). So as not to pollute the major cities in the southwest with air and water contaminants, the power-producing plants were located in unpopulated desert regions in Nevada, New Mexico, Utah, and Arizona. Yet the first plant to be built (near Farmington, New Mexico) threw out effluent that could be tracked over 200 miles away by aerial observers, and from 170 miles high by the Gemini 12 astronauts. When the entire complex is completed, it is estimated that it will emit nearly 2,000 tons of sulfur dioxide, nearly 1,300 tons of nitrogen oxides, and 240 tons of fly ash each day, degrading the heretofore clear desert skies and polluting six national parks, twenty-eight national monuments, and three national recreational areas—all this in spite of the installation of antipollution devices!

Observations from ERTS 1, Skylab, and other orbiting spacecraft show the relationship between national wilderness and recreational areas, the offending power plants and the cities they serve. They also show as well as permit the mon-

Distribution and dispersion of water pollutants are observed with overview imagery. Here, at the mouth of the Maumee River at Lake Erie, a sewage disposal plant (1) is seen along with the sewage it discharges into the river (2). The breakwater (3) impedes circulation of the effluents into the Lake. (NASA, U.S. DEPT. OF INTERIOR/GEOLOGICAL SURVEY)

itoring of the Colorado River that has to supply some quarter of a million acre feet of water each year for the steam generators associated with the power generation complex, putting a further strain on downriver farming activities already hit by increasing salinity. Worse still, a strip mine at Black Mesa, Arizona, must supply, via a coal-water slurry sluice, the coal needed by the Navajo plant near Page, Arizona. The water has to come from somewhere, and that turns out to be the underground reservoirs beneath lands

inhabited by Navajo and Hopi Indians, whose wells have suffered. Despite the fact that the mining companies offered millions of dollars in royalties and wages each year, one Indian lamented: "We don't want money from coal companies. We love all earth and all nature. We get life from earth. We get food. Money go away fast. Then you have nothing left. We have no land, then we have nothing."

Situations like this face geographers and environmentalists all over the world, from the huge Aswan Dam to raising the level of Ross Dam at the request of the Seattle City Light Company. The ecological impact of the Egyptian dam is extremely complex and may end up causing more environmental damage and human suffering than even the most pessimistic observers now imagine. As for the Ross project, examination shows that raising it will flood the Big Beaver valley, destroying large stands of cedar and numerous beaver dams and ponds and drowning a major trout stream in the Vancouver area.

Highways, power installations, urban centers all reflect the nonagricultural use of land by man. The size of the continents and islands being constant, in a world of

Lake Nasser, the immense body of water backed by Aswan Dam, curls through the center of this Apollo 9 photograph of Egypt. On the horizon is the Red Sea. (NASA)

exploding populations and prodigious human activity, available land inevitably becomes scarcer and more expensive. Yet vast amounts of land are required for farming and raising livestock. Still other land is needed to support the forests that provide wood for construction, paper-making, and other essentials of a technological society. The ability of land to produce optimum crops, animals, and trees depends partly on natural factors and partly on how humans go about conserving, irrigating, and enriching it. And again, the satellite overview is immensely useful.

For example, drought conditions were exceptionally severe in the southwestern United States in 1971. On the Texas–Oklahoma border, the Red River went virtually dry, the surrounding area receiving only an inch of rain during the year as opposed to the normal twenty-seven inches. Only a tenth to a twentieth of the normal production of wheat was realized in many parts of the southwest and animals perished by the tens of thousands. In the spring of 1973, on the other hand, large areas of the southeastern part of the nation reeled under the impact of abnormal quantities of rain which led to widespread flooding of the Mississippi, Tennessee, and other river systems. Shortly after, severe tornados arrived on the scene to create further destruction. Not only was property damage extensive and millions of trees destroyed, but vast amounts of fertile cropland suffered at a particularly vulnerable point in the planting and growing cycle.

The results of such conditions, and, indeed, their very spread, can be observed from orbit. Based on day-by-day receipts of information on both local and regional conditions, corrective actions can be

planned—or, at the very least, decisions made—as how best to face up to a tragedy. Pockets of cattle dying of thirst may be located, areas defined where withered crops should quickly be plowed under, and the advance of such dry-weather thriving growths as mesquite shrub observed. Conversely, areas of actual and potential flooding can be quickly and regularly observed. The advantages of the overview to land-use planners is expanding all the time, as information is diffused on the capabilities of orbital systems. In a highly populated state like Ohio, particular interest has been exhibited for land-use map preparation, a variety of cartographic applications, and thematic mapping (for example, census and transportation). Members of the State of Ohio Department of Development and the Battelle Columbus Laboratories are updating the state's land-use survey of 1960, using photography from ERTS and Skylab to support experimental preparation of base maps, topographical maps, photomosaics, and other special-purpose maps for demographic, urban development, and transportation interests. They are also applying orbital survey capabilities to controversial strip-mining reclamation efforts in Ohio. In an assessment of the importance of the orbital overview for the state, G. E. Wukelic, T. L. Wells, and B. R. Brace reported:

Most significant short-term benefits are considered possible in the categories of legislation and state government reorganization. The Ohio Legislature is considering numerous natural resources and environment bills, the development, implementation, and enforcement of which could be heavily influenced by automated and manned

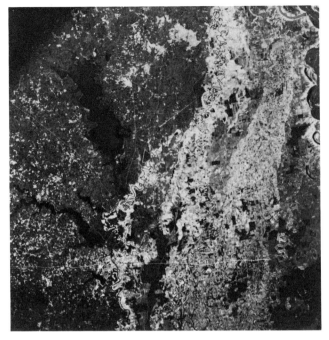

Monitoring floods and surface water changes: Apollo 9 photograph (left) of Louisiana in March 1969 shows the Ouachita River (upper left) in full flood, covering about 165 square miles; Mississippi River is at extreme right. Below, ERTS 1 photographs recorded marked changes in Susanville, California, reservoirs over an eighteen-day period in the summer of 1972. In the first photo (middle), made on July 25, the reservoirs are about the same size as on the map (left); in the second photo, taken on August 13, McCoy Flat Reservoir (1) has nearly dried up, while Mt. Meadows Reservoir (2) has decreased about 50 percent in area. (NASA, U.S. DEPT. OF INTERIOR/GEOLOGICAL SURVEY)

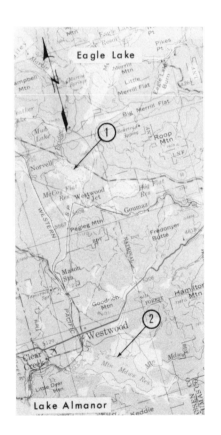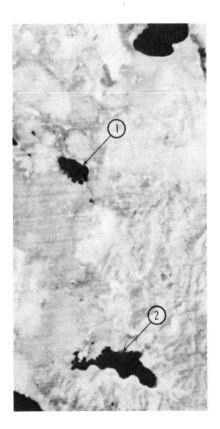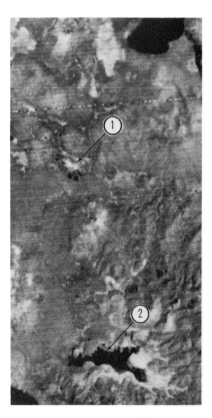

43

satellite capabilities. For example, the distribution of appropriate satellite-acquired photographs could provide broader perspective on environmental issues requiring legislation. Likewise, Ohio's ERTS and Skylab experiences and findings should prove valuable to studies in progress concerned with state reorganizational possibilities. This will be especially so for considerations regarding how the state should be reorganized to be most responsive to increasing resource and environmental issues, the delivery of state services, and associated federal controls and funding opportunities.

On a longer-term basis, we anticipate benefits to accrue from operational orbital surveys which will impact on all Ohio resource- and environment-related problem areas. However, most important are those potential benefits anticipated in the budget, development, and employment categories. Certainly, even if only partially successful, anticipated cost savings inherent in operational orbital surveys will make many new and necessary resource and environmental programs possible. A large percentage of these are currently being rejected solely on economic grounds.

Air and water quality regulations and natural gas shortages pose serious national, industrial, and community development problems. Repetitive orbital survey data could be quite useful in long-term planning of types and locations of new industries and new towns in Ohio. A technically sound and positive attitude toward planned industrial expansion is essential to maintaining a healthy economy in Ohio as well as to improving the unemployment situation, both of which will worsen if unreasonable environmental restrictions are imposed.

Means of using satellites for demographic research are only now being worked out, in Ohio, Rhode Island, and elsewhere. Certainly, repetitive observations will permit urban buildup and the expansion of suburbs to be monitored, and the appearance of slum-cleared areas characteristic of many central cities to be observed. Demographers estimate that it took centuries for man to propagate to his first billion, but only eighty years his second, and less than forty years to reach the 1974 world population level of 3.9 billion. Estimates for the year 2000 are 7 billion; and, for A.D. 2050 30 billion, based on a world population increase of about 2 percent per year. Population rise has, of course, an immediate impact on land: the United States, with less than 6 percent of the world's people, consumes about 40 percent of the world's resources—and in the process discards over a million motor vehicles, 36 billion bottles, and 58 million tons of paper each year. Paul Ehrlich once said, "America's pride in her growing population may be compared to a cancer patient's pride in his growing tumor." Yet population growth alone fails to tell the whole story of man's deleterious effect on geography, for, as Samuel Preston of the University of California at Berkeley cautions, "When you look at demands on natural resources, two-thirds of that is because of per capita growth and one-third because of population growth." Barry Commoner of Washington University supports this view, asserting that since 1946 the population of the United States rose by 42 percent compared with a 2000 percent increase in pollution.

44

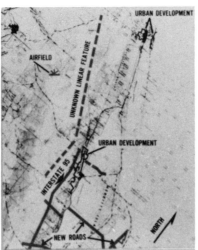

Major cultural changes identified by Gemini 7 photography of the Cape Canaveral area of central Florida. Left, orbital photo taken in December 1965. Right, map overlay of principal man-made changes to area. (NASA)

Our renaissance of awareness [of such facts as these] has come none too soon," in the opinion of Harlan Hatcher of the University of Michigan, "and we hope also not too late. What we humans have done to our only habitation begins to shame us. What we are still doing to it begins to alarm us. What we do with it and to it in the future under continuous population pressure and industrialization has now become a priority concern of critical importance. And our time scale and margins, our target dates, are no longer leisurely millenniums as in the past, but brief decades or at the most and with good fortune a few generations. Our unflagging purpose must be to leave the earth in no worse shape than when we so briefly inherited it and had responsibility for it.

If major geographically related dilemmas (the destruction of the environment, the covering and the modification of the land, the depletion of vital petroleum, solid mineral, and other resources, and an exploding population that daily exacerbates all the other problems) are to be met and at least contained, regular, accurate, up-to-date and complete information must be available on a worldwide basis. Continuous global examination of the state of man's works and activities may well lead to the conclusion that his aspirations for an ever higher quality of life are doomed to failure unless drastic changes are made to his ability to control population growth and the urge of the existing population to cover the Earth with asphalt and garbage.

More than ever before, we realize that our Earth is a great life-support system, probably the only one we can count on. We know that what man does in one corner of the world today can have important, often disastrous effects on another tomorrow. This has led a well-known ecologist, Garret Hardin of the University of California at Santa Barbara, to say of the nation's original investment in the Apollo manned lunar exploration programs:

if [the space program] results in our really believing in our bones that our earth is a spaceship, and the only one we will ever continuously inhabit for a long time (and perhaps forever), then twenty-four billion dollars is a cheap price to pay for so essential a lesson.

45

Snow-crowned peaks of the Hindu Kush range in the Himalayas from Apollo 9. (NASA)

# 3

# Down-to-Earth
# Geology from Space

CAPCOM: *Oh, and Apollo 9, I have about three more targets here we'd like [you] to photograph. One of them is coming up in about 7 or 8 minutes. If you can't make it, well, no sweat.*

SPACECRAFT: *Go ahead.*

CAPCOM: *OK, the first one. 212 plus 04 plus 16, four frames, 7-second intervals. Zero degrees; this is of Morocco, for geology. . . .*

SPACECRAFT: *Stu, we've got it.*

CAPCOM: *Oh, OK. The second one is . . . weather time, 2121056, four frames, 6-second exposure intervals, zero degrees. And these are the old Tibesti Mountains here in Chad, and you are going to come up over them this time, and our first one is . . . seven pictures at 19-second intervals, zero degrees. And this is with Ethiopia Rift Valley, studying up on the geology there, and the last one is . . . three pictures, 8-second intervals . . . and this is geology and this is of Somalia.*

—APOLLO 9, MARCH 3, 1969

The need to expand natural resource reserves in the face of rapidly rising demands has caused the pace of geological research to accelerate rapidly since the early 1960s. Geologists are particularly intrigued with the exciting information available from the satellite overview, which is already helping them to reduce the costs of—and increase the yields from —the search for mineral fuels, metallic ores, and nonmetallic deposits of many kinds.

The geologic observations of the Apollo 9 astronauts were part of the Earth Resources Observations Satellite (later Systems) Program, established by the then Secretary of the Interior Stewart L. Udall on September 21, 1966. Commonly known by the acronym EROS, it sought to make optimum use of information generated from aircraft as well as spacecraft for the United States Geological Survey and other Interior bureaus, including Mines, Land Management, Reclamation, Commercial Fisheries, Outdoor Recreation, the National Park Service, and Indian Affairs. To handle and disseminate land area imagery from the ERTS 1 spacecraft, from the Earth Resources Experiment Package aboard the Skylab space station, and from other sources, the EROS Data Center was set up by the Geological Survey at Sioux Falls, S.D. (oceanographic imagery is handled by the Department of Commerce, Suitland, Md.).

Some of the things to be gained by geologists from satellite observations were cited in a 1966 report by scientists at the Infrared and Optical Sensor Laboratory, University of Michigan:

Remote sensors are a very important aid in exploring the earth's crust and will no doubt be instrumental in further exploring presently productive regions for mineral and petroleum and locating new wealth in unmapped portions of the world . . . Remote sensing techniques can be used to locate areas favorable to mineral and petroleum accumulations by the identification of geologic features peculiar to such areas. This preliminary reconnaissance makes it possible to concentrate field investigation in relatively limited areas. Spacecraft sensors could readily detect many types of geologic features; once located, any number of anomalies could be examined in greater detail by remote sensing from aircraft and even more limited areas would be investigated by field parties.

Geologists realized that since it was no longer common to discover exposed deposits of valuable ores, new methods had to be employed to detect and assess structural and lithologic conditions favoring the occurrence of hidden but potentially extractable resources. They also knew that repetitive observation is necessary (even though geological features are the least changeable of all Earth resources) because different viewing conditions often give rise to more accurate interpretations. Thus, a given geologic province or feature appears different when lit from a number of angles; when covered by differing stages of vegetation growth and by snow; and when wet, moist, and dry.

The new EROS quickly began showing its worth. One of the early projects was the preparation by the Geological Survey of a photo-image map of Peru and portions of neighboring Bolivia and Chile based on twelve rectified Gemini photos and displayed on a scale of approximately 1:1,000,000. William A. Fischer of the Survey reported that a third of a million square miles were photographed in less than three minutes in June 1966 by Gemini 9 astronauts Eugene A. Cernan and Thomas P. Stafford, and that the cost of compilation was about a tenth of a cent per square mile. He found that

Landforms are portrayed in greater detail on the photo-image map than is possible to show with conventional contours on small scale maps. Further, the photo-map contains a great deal of information that is not commonly shown on [such] maps. For example, one may see distributions of gross rock types, present and prior levels of lakes, salt flats, waterfalls, mining areas, roads, irrigated lands, jungle agricultural sites,

The rugged west coast of South America from Apollo 7. In the foreground rises the 22,000-foot-high Llullaillaco Volcano. At center, behind the large salt lake and atop a 19,000-foot-high volcano, the countries of Bolivia, Argentina, and Chile meet at a common point. Below the clouds in upper portion of picture are the great Gran Chaco plains. (NASA)

landslides, grasslands, snow, archeological sites, smoke and smog . . . two circular structures are visible that, to the best of our knowledge, have not been previously reported and were not recognized on the unrectified photographs.

Each photograph was essentially orthographic, meaning that features were seen in their true relative geographic positions in spite of elevation differences. As an example of a major geologic discovery made from the mosaic, an important fault zone was traced over seven hundred miles through the Andes mountains from a point near Toquepala practically to Cerro de Pasco. Another geologist W. D. Carter, reported that additional ". . . lineaments and major folds [were] seen that would take years to photograph from aircraft and decades to map on the ground, if they could be recognized at all."

Carter estimated that the potential benefits of the space overview to Department of Interior activities in the United States would be on the order of $80 million a year, and that benefits to the private sector ". . . appear to be many times that level." Citing a single example, he stated that ". . . better knowledge of soil moisture distribution could reduce the hours of irrigation, thereby saving water in arid lands and reducing the hazard of disturbing soil chemistry." Some of the "most obvious" applications of space sensing for a typical mineral exploration company are, according to Carter:

1. As each photograph will be orthographic, in that all features will be in their true geographic position, they will serve as photographic base maps similar to our 2-degree sheets at a scale of 1:250,000; topographic contours can be overprinted if desired; they will be up-to-date.

2. They will show all major geologic structures expressed by topography or soil moisture in their true relationship and without human bias.

3. Intrusive masses, alteration zones, mine workings, dumps and access roads, if large enough and of sufficient contrast, should be visible so that mining districts, mineral belts, or metallogenic provinces can be plotted directly on single photographs or a mosaic of a few overlapping photographs.

4. The distribution of moisture on the land surface should lead to the discovery of unmapped faults and fractures of extensions of known structural features that, in places, may be associated with mineral deposits.

5. Repeated observation of plants and their conditions of vigor throughout the year should reveal anomalous conditions; most will probably be due to disease or insect infestation but some will undoubtedly be due to chemical conditions of the soil. Some of these may eventually lead to the identification of areas overlying mineral deposits.

6. Repeated observations of the continental shelf and river effluents will help us to map and understand the effects of currents and tides on the distribution of sediments and could lead to the location of sub-aqueous placer deposits of value.

7. Repeated observations on a seasonal basis of such features as snow fall and rates of melt may help us to find alteration zones where sulfides are actively oxidizing at or near the surface.

8. Repeated observations of active geologic processes such as volcanic erup-

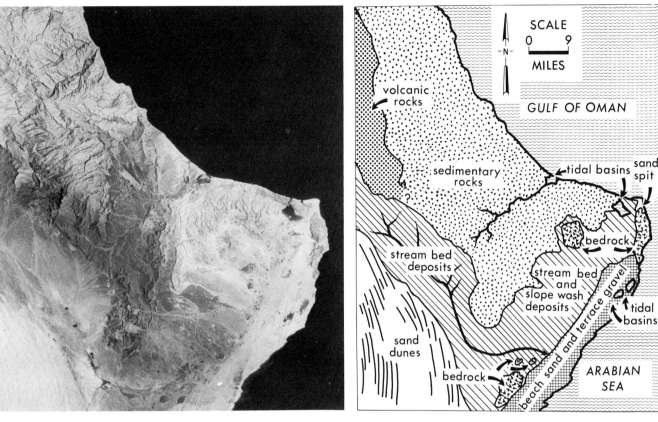

Ras al Hadd, prominent cape at the southeastern tip of Saudi Arabia, taken from nearly straight down by Gemini 4. At top is Gulf of Oman; at bottom is Arabian Sea; at lower left are Seif Dunes in "Empty Quarter." A geological analysis of the area is at right. (NASA, U.S. DEPT. OF INTERIOR/GEOLOGICAL SURVEY)

tions, earthquakes, hot springs, landslides, avalanches, and floods should help us to map and understand the processes and improve our techniques of disaster warning. Better understanding of active geologic processes should improve our ability to find ore deposits.

Exploration geologists working with the Geological Survey's Geologic Applications Program utilize to the fullest space-generated data to improve the speed, accuracy, and perspective of both geologic mapping and resource evaluation. At first, they used photography from the manned Mercury and Gemini programs as well as

such unmanned craft as the Nimbus series, finding distinct advantages to the synoptic views obtained under uniform lighting. They also learned that the angle of illumination of the ground was very important to the correct interpretation of major geologic features (faults, fold belts, etc.), which for the first time could be seen in their full regional context; moreover, much light was shed on their connection with intrusive masses and known mineral beds. In one infrared survey experiment over the Carrizo plains area of the San Andreas Fault zone, imagery helped reveal soil moisture and ground water entrapment in places where

51

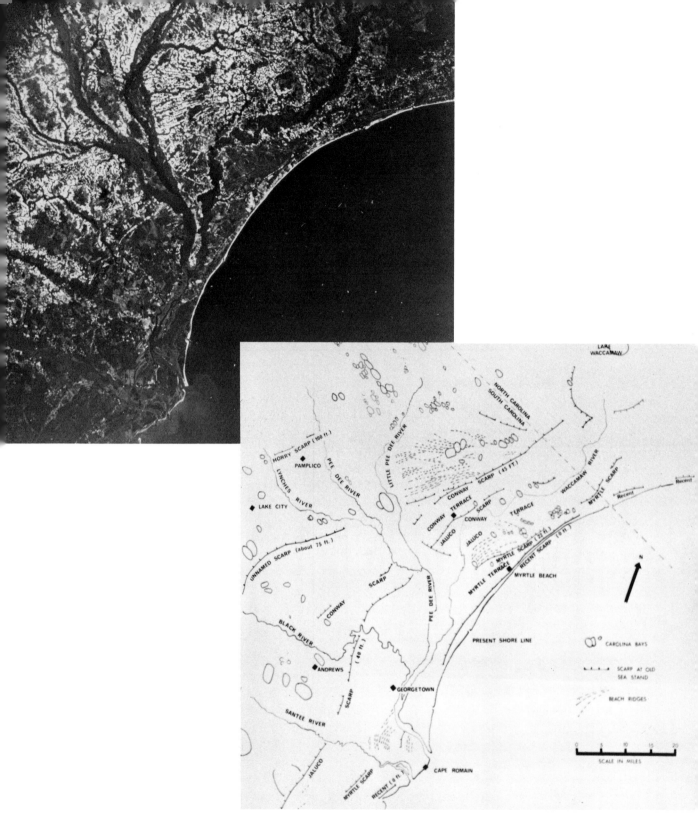

Light snow in Apollo photograph at top of the Georgetown, South Carolina, area enhances small topographic features. Above the many fine lines, running generally parallel to the modern coast, are ancient beach ridges and shoreline scarps. (NASA, U.S. DEPT. OF INTERIOR/GEOLOGICAL SURVEY)

alluvium covered the fault. Also, thermal differences were detected in both compact and loose rock and soils during the twenty-four-hour day-night cycle.

Another early project of the Geological Survey, based on Gemini photographs, was the publication of a small-scale photo-map of more than one million square miles of the Earth's surface at a cost of less than $25,000—a small fraction of what the same job would have cost using conventional techniques. The Survey also published a tectonic map of the south-western United States, useful in regional exploration for fuel and mineral deposits, and a geologic terrain map showing the distribution of superficial deposits in the southwest and in northern Mexico.

Following the last of the Gemini flights geologic photography became an integral part of the Apollo program. In the unmanned Apollo 6 flight in April 1968, geologists learned how important low-Sun-angle orbital photography was in areas of good visibility and high topographic relief, and that overlapping vertical photography from space was extremely valuable in assuring continuous coverage of large geologic areas as well as stereoscopic study without rectification. The flight not only clarified the existence and the nature of a major geologic feature, the Texas lineament, but permitted refinements to be made to geologic maps of Baja California and New Mexico.

Geologic studies based on photographs from the October 1968 manned Apollo 7 mission were made of such areas as Baja California, the Middle East, southeast Brazil, and Iran. As a result of photography made at various Sun angles, altitudes, and oblique views, structural conditions were revealed in the highlands around the Red Sea and the Gulf of Aqaba that

helped solve the origin of the African rift valleys. Malcolm M. Clark, in the course of interpreting Apollo 7 photographs, remarked that they ". . . are useful in geology because they reveal features of such extent, subtelty, or discontinuity that the features become evident only at the small scales obtainable from orbit."

Another means of employing the overview to geological phenomena involves the detection of color patterns and their association with mineral resources. Geologists know that some bright-colored surface weathering effects are produced above valuable primary ore deposits, though D. R. Leuder cautions that color itself is not a "localizer" but rather only an "indicator." "The color," he writes, "may be associated with the actual outcrop, an oxidization stain indicative of underlying minerals, or an alteration halo. In any case, the blacks, reds, greens, browns, and yellows associated with mineralization always bear investigation." The modern way to do this is from above.

Mineral deposits also may be indicated by plant colors and associations. For example, the appearance of foliage may be affected by minerals that have dissolved in the soil water and been absorbed by the roots. Moreover, some plants enjoy unusual vigor while others wilt or even die in the presence of certain minerals. For instance, luxuriant ragweed may indicate zinc; brilliant red campions suggest copper. In this connection, F. C. Canney, of the U.S. Geological Survey states that "Traditional prospecting methods, wherein rocks are examined directly for valuable minerals, can no longer be expected to produce many important mineral discoveries, except possibly for those deposits of minerals that were of little or no economic interest until recently." Many large

53

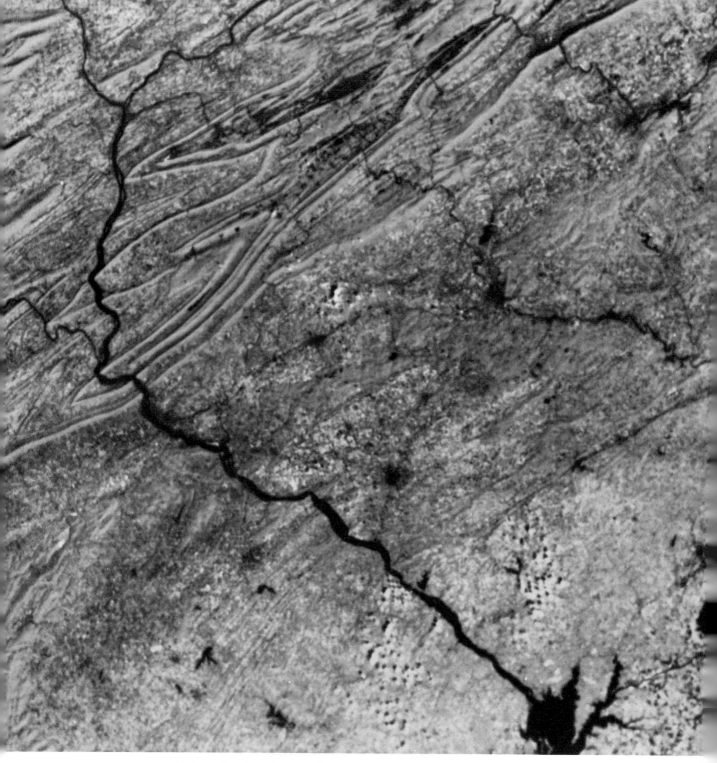

The Allegheny Mountains of southeastern Pennsylvania stand out in sharp relief in the infrared of this ERTS 1 photo. The Susquehanna River runs from upper left to lower right; the Delaware River and Wilmington are in the extreme lower right. Harrisburg is left of center; Lancaster, just below center, and Reading, right of center. (NASA)

A near vertical view from Skylab 4 of the snow-covered northwest corner of Wyoming. Dark area is Yellowstone National Park; largest body of water is Yellowstone Lake. Mountain ranges include the Absaroka (east and northeast of Yellowstone Lake), the Big Horn Mountains (eastern part of photo), and the Wind River Range (bottom center). (NASA)

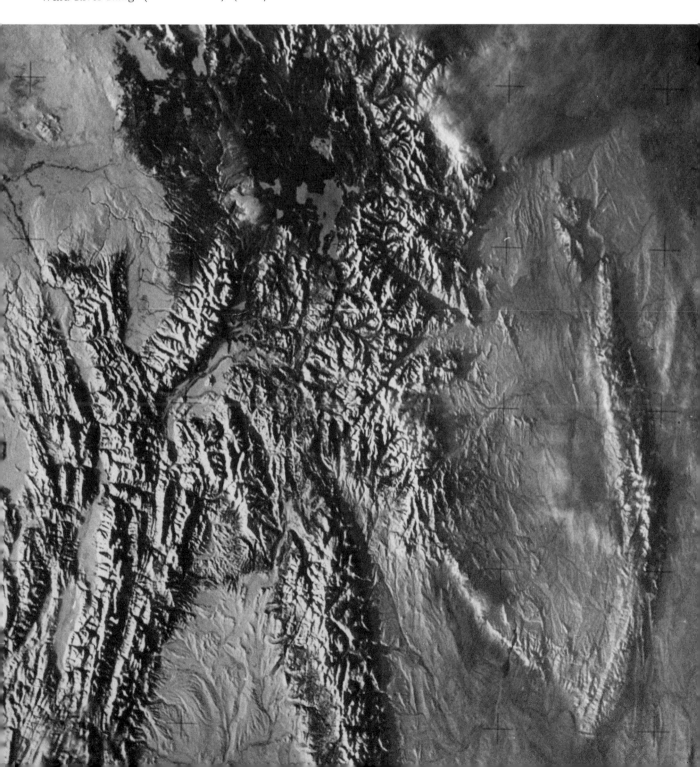

areas, rocks and geologic structures favorable for the occurrence of ore often lie "concealed beneath soil and alluvium." In order to discover and then delineate metal anomalies in soils, hundreds if not thousands of soil samples must be taken and then analyzed at very high cost. Once again, the overhead view seems to have much to offer. Canney explains:

In considering ways of sensing abnormal chemical conditions in the soil by remote means, the possible use of vegetation is, for two reasons, a natural avenue to explore. First, data from many biogeochemical surveys performed during the past few decades have shown that plants growing in a geochemically anomalous soil generally reflect this in their trace element content; and, sometimes, these plants show characteristic variations in their form, color, size, or

rate of growth. Second, the forest canopy is easily visible to a sensor in a plane or a satellite.

The program of geologic photography has continued through the Apollo 9, ERTS 1, and several Skylab missions and many new discoveries have been made. Not long after ERTS 1 was launched, for example, previously unmapped geologic features along California's central coast range were located—and this despite years of intensive study of the area. To the east, in Mississippi, previously unknown salt domes were discovered. And in Alaska, to the west of the Prudhoe Bay oil fields and north of the settlement of Umiat, a possible petroleum-bearing structure was found. Geologists of the U.S. Geological Survey suspect that the elongation of individual lakes and their alignment around an elliptical structure seen

New ground details revealed through space photography: Left, this Tiros 9 photograph led to the discovery of two previously unknown faults in northern Norway; they extend southward into Sweden, where they bound the Skelleftea iron ore deposit. Right, ERTS 1 photography defined oil exploration targets in northern Alaska. Key features, indicated by letters, include: (A) elongation of lakes, confirming a known linear trend; (B) linear alignment of lakes, suggesting a newly defined trend, and (C) curvilinear clustering of lake, suggestive of a subsurface structure. (NASA/ESSA AND U.S. DEPT. OF INTERIOR/GEOLOGICAL SURVEY)

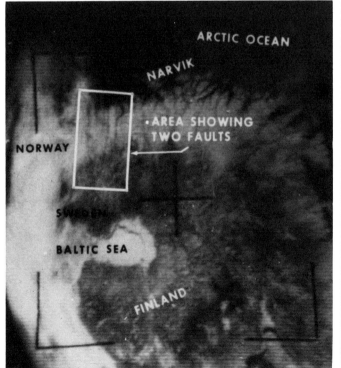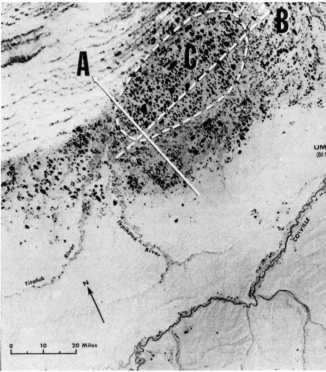

on ERTS imagery may well be associated with oil or gas-producing formations, or both.

As so often happens, advances in one area carry penalties in others. When ERTS and Skylab surveys help ground-based field teams locate economically attractive deposits beneath plant and soil cover, the danger of environmentally ruinous strip mining immediately arises. Strip mining already accounts for two-fifths of the coal produced in the United States as well as significant percentages of other resources, including iron and copper. Coal companies estimate that it costs about $1.50 a ton less to strip-mine than to go underground, and therefore greatly favor the practice. It is possible to restore strip-mined areas, but the costs can be high—in some cases so high as to discourage the act in the first place if laws requiring companies to restore the land were on the books—and if the laws were enforced. In Pennsylvania much scoured land has been backfilled to the original contours, with highwalls and spill banks removed. Moreover, it is done as quickly as possible after the coal is mined so that acid-forming strata are exposed for the shortest time possible, reducing water pollution. What polluted water that does collect is often treated immediately in settlement basins.

The effects of strip mines on water pollution have come to the attention of ERTS 1 experimenters, notably Wayne Pettyjon of the Ohio State University. "Mineralization," he notes, "increases astronomically [in streams running through strip-mined areas]. Sulphase and iron become very high and very few types of aquatic life can survive. . . . One of the big problems is that we don't know how much area on a day-to-day, month-to-month, or year-to-

year basis has been stripped. In other cases, we don't know where some of the old mines are. . . " He and his coworkers employ ERTS 1 photography to map areas that were mined as far back as 1935, helping them evaluate the recovery of vegetation. "I suspect," he adds, "that we will find some areas where vegetation can take over naturally and others where normal reforestation techniques won't work."

Few states have effective strip-mine restoration laws, but eventually strong controls must be exercised throughout the U.S. and other major mining countries. Once again, ERTS and Skylab have already played an important role in monitoring strip-mining activities and in spotting areas where restoration is proceeding slowly or is not being done at all. The geologist is unable to restore natural resources in the manner that the forester can, but at least he looks forward to the time when all mining scars will be healed and the lands returned to their original appearance. This, alone, will be an important dividend from space.

In addition to the mapping of major structures such as folds, faults, lineaments, intrusive masses, and mineral belts, much attention is being focused in the geological phases of ERTS and Skylab on the detection and monitoring of thermal activities, including those associated with volcanoes. Geologists believe that by regularly examining so-called thermal anomalies, it may be possible to provide warnings of impending volcanic eruptions. For example, satellite observations of out-of-season melting snows on the flanks of volcanoes during the winter months may indicate abnormal heat flow—and thus, activity—on the inside. Counts of local seismic events also have

Strip-mining scar: The light, circular spot indicated by arrow is an 80-mile square desertlike area, the result of a copper mining operation that wiped out all the vegetation in what had been a fertile Appalachian valley. This **Apollo 9** view looks north from Georgia, toward North Carolina, Kentucky, eastern Tennessee, and Virginia. (NASA)

Analysis of fracture pattern, such as in this near-vertical Gemini 4 view of southwestern Saudi Arabia, helps geologists to search for underlying domal structures that may be traps for oil. At upper right are the complex longitudinal dunes of the Empty Quarter. To compare the differences between color and black and white reproduction, see the original color version of this photograph on page 86. (NASA)

Dark vertical bands in lower left of this Skylab 2 photograph of southeastern Utah outline the San Rafael Swell, a large upwelling of the Earth, called an anticlinal structure. Such structures often are associated with mineral deposits, including oil and gas. (NASA)

been made, changes being noted in the inclination of the volcano surface and measurements being made of temperatures and gas composition in fumaroles.

As part of the ERTS 1 program, some fifteen volcanoes from Alaska southward to Central America were monitored; and, in one instance in Guatemala, an increase in seismic activity was observed to precede

an eruption that took place in February 1973. From an average of five seismic events per day prior to the buildup toward eruption the rate climbed just two days before to eighty counts per day. Meanwhile, in the North Atlantic the NOAA 2 satellite measured not only the aerial extent of an eruption that took place on the island of Heimaey off the

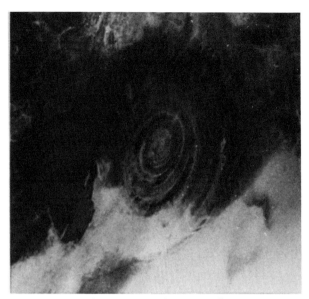

Crater Richat in the Sahara Desert. Mosaic at left is not as revealing as the single Gemini 4 photo at right, which shows deformation in the surrounding bedrock. The circular feature, thirty miles in diameter, has been interpreted as either a meteor impact area or the result of an igneous plug pushing up from great depths. (NASA, U.S. DEPT. OF INTERIOR/GEOLOGICAL SURVEY)

southern coast of Iceland but its thermal radiation as well.

As a by-product of these studies, new geothermal sources may be revealed. Infrared surveys over the Norris Geyser Basin in Yellowstone Park showed how thermal anomalies were distributed, helping geologists establish guidelines for locating geothermal power sources. It is also known that by watching the melting patterns of snow, the existence of undiscovered mineral deposits may be inferred. Some deposits produce heat from the process of oxidization, which often affects the melting rates of overlying snow-cover. Also, overview infrared surveys can lead to the detection of underground coal fires that produce lethal, upward-seeping gases.

Earth resources satellites also may be used in the future to help forecast earthquakes. L.C. Pakiser and his co-workers at the Geological Survey's National Center for Earthquake Research, Menlo Park, California, have reported that "The rate of movement along different segments of the San Andreas Fault in seismically active areas was observed to change before the occurrence of moderate earthquakes," leading them to suppose that it ". . . seems reasonable to hope that short-range prediction of earthquakes (on the order of hours or days) may be achieved through the *continuous* monitoring of ground tilt, strain, seismic activity, and possibly fluctuations in the earth's magnetic field." Many geologists suggest placing strain gauges, tiltmeters and other devices in myriad locations around earthquake zones (including remote islands and on the Antarctic continent) and then monitoring them regularly by satellites. These orbiting vehicles would instantaneously relay the information to ground interpretation centers, where predictions of impending dangers would be made. There seems to be no economic substitute for the satellite in this application.

61

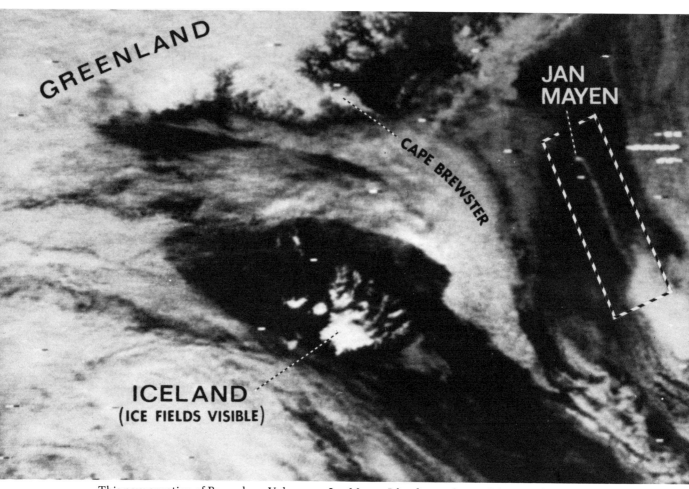

**GREENLAND**

**CAPE BREWSTER**

**JAN MAYEN**

**ICELAND**
**(ICE FIELDS VISIBLE)**

This new eruption of Beerenberg Volcano on Jan Mayen Island was observed first on the night of September 20, 1970. By noon of the next day, when Nimbus 4 took this picture, the ash plume, within the dotted rectangle, extended more than 200 miles to the southeast. (NASA)

ERTS 1 was not programmed to observe the Managua, Nicaragua, area prior to the devastating earthquake of December 23, 1972, but it began to gather data almost immediately afterward. In concert with an Earth resources aircraft dispatched from NASA's Johnson Space Center, the satellite yielded sufficient information to allow a chart to be prepared delineating faults and other linear fea-tures related to the earthquake. Geologic interpretations were later turned over to the Nicaraguan government to assist in planning for the reconstruction of the capital city.

Despite many encouraging experiments made by NASA, the U.S. Geological Survey, other national and foreign agencies, companies and universities, geologic satellite technology is still in the develop-

The San Andreas Fault is revealed in pristine detail in this radar image of the San Francisco Peninsula. This image was made by airborne equipment in 1965 as part of NASA's program to develop satellite sensors. (NASA)

ment stage and as such the economic benefits it eventually will make possible are difficult to discern with accuracy. But we do know that the U.S. is a prodigious consumer of natural resources and that it is predicted to double its present rate of consumption within the next two decades. Even though it is richly endowed with its own natural resources, its industries require more and more imported materials every year, particularly petroleum, implying an interest in increasing not only its own internal reserves but those of less industrialized countries on which it is dependent. The winter 1973-1974 world oil crisis triggered by the Arab oil embargo against many industrialized nations graphically demonstrated to the United States the need to strengthen its own petroleum resources. Skylab, ERTS, and other overview imagery is an important and economical tool to meet this objective.

Geologic and other orbital overview imagery can be processed by computer, leading to what is called computer image enhancement. In February 1974, Alexander Goetz and Fred Billingsley of the Jet

63

Propulsion Laboratory—California Institute of Technology reported the successful use of such enhancements to identify surface manifestations of mineral deposits near Goldfield, Nevada. "We have proved the ability of the ERTS imaging system to detect known mineralized zones," said Goetz, "and we hope to extrapolate our findings to other areas . . . Ores cannot be detected directly by the ERTS cameras. They cannot see beneath the top millimeter of the Earth's crust," he added. "What we look for are secondary indications such as hydrothermal alteration products (minerals which formed on the surface by long geological seepage). These include iron oxides and clays which are orange or white in color but which have characteristic spectral signatures in the infrared." Noting that in computer-enhanced ERTS photos the mineral deposits appear green or brownish green, Goetz reported that some 20 different points observed by the spacecraft's cameras were verified to have been mined already or "to have gold or silver potential in a 3,000 square mile area east and south of Goldfield. There is a distinct correlation between our map and known ground findings."

The United States and Canada spend

well over $600 million each year on geologic, geophysical, and mapping operations directly related to the search for oil and mineral deposits, including $345 million for petroleum in the U.S. and $35 million in Canada, and $200 million and $50 million in the two countries for mining. Another $65 million are spent in preparing regional geologic maps and in carrying out broad geophysical studies. If the data collected as a result of geologic and geophysical exploration activities are useful for only one decade, even a 1 percent improvement in efficiency by using geologic satellites would make the system most attractive (yielding well over $60 million). Yet the United States Geological Survey has estimated that such a system will produce an improvement in exploration efficiency of 7 percent. Whatever the accuracy may be of this particular estimate, there is little doubt that the use of remote geologic sensors is stimulating the discovery of new and valuable oil and mineral reserves.

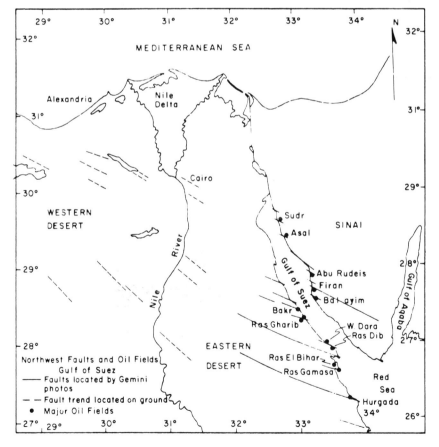

The whole of Sinai Peninsula, as seen from Gemini II together with geological sketch map, showing minor oil fields and northwest faults. (NASA)

65

Lake Ontario, Canada above, and New York below; the eastern tip of Lake Erie; and the Niagara Falls area between the lakes, as seen from Skylab 3 in 1973. (NASA)

# 4

# Inland Waters, Snow, and Ice

The land shines with waters—inland seas, lakes, ponds, marshes; it glistens with ice and snow. Water flows to the oceans in streams and rivers but much also is trapped permanently in soil and rock interstices beneath the surface. For ages, the availability of water determined where man lived—neither cave nor city could be far from life-giving water—and some of his most impressive feats of engineering have been inspired by the need to contain, restrain, conserve, and channel the waters of planet Earth.

The waters on and beneath the land derive ultimately from the oceans. Water evaporates from them, is transported by the winds, and falls as rain, snow, or hail. About two-thirds of land waters go back into the atmosphere by evaporation or plant transpiration with the remainder returning directly to the ocean by surface flow or underground seepage.

The science of hydrology seeks not only to describe and understand more fully the Sun-generated water cycle, but to predict how the surface and ground waters will

67

move and accumulate, and to manage them to meet man's needs. One of the hydrologist's major goals is to even out the supply—to plan dams and reservoirs that will at once prevent floods and mitigate droughts. He brings water to dry and desert areas by diverting it from distant sources, by building local reservoirs, or by drilling wells. In areas of heavier precipitation, he deals with problems of soil erosion, water pollution, inland waterway transportation, and the expansion of water power to create electricity. Accumulations of ice and snow concern him, and he may spend much of the winter trying to determine the probable effects of the spring melt for good (irrigation) and for bad (floods).

As populations expand, as man withdraws more and more land from its natural state, and as the quality of water deteriorates due to the hazards of pollution, hydrologists find themselves severely pressed to meet their objectives and their growing responsibilities to the well-being of society. Fortunately, the advent of unmanned and manned satellites portends significant advances in hydrology at the very time when the hydrologist needs them most. Both theoretical and applied aspects of hydrology depend to a great extent on the regular and accurate receipt of data from stations scattered across vast geographical expanses, many of them in remote and inaccessible areas. Some data are directly obtained either by manned or remote sensor observation, while others are measured on the surface by automatic instruments (such as stream gauges) that transmit their hydrologic outputs to data relay satellites.

Photographs from orbit provide invaluable information on the extent, distribution, and density of river channels; the size and shape of lakes and reservoirs; the buildup, migration, and dissipation of pollutants, and the specialized ground features that control the storage and movements of water. Nonphotographic sensors help the hydrologist assess, on a repetitive basis, the heating, cooling, and evaporation patterns of water bodies and the general delineation of drainage basins. They also give him insights into ground water distribution as expressed by vegetation vigor, the relative rates that rocks and soils dry out, and contribution of ground water to surface water bodies and the influence of the latter upon the former. By constructing accurate, up-to-date maps of water table levels and changes, useful information can be gained on progressive changes to area ecology. Infrared devices also can detect geologic faults and fractures that affect the movement of ground water. Since the estimated capital investment in U.S. water resources is in excess of $235 billion and is increasing at the rate of $7.5 billion a year, economic considerations alone dictate that ERTS and other satellite technology be vigorously applied to the vital field of hydrology.

Although major water systems often transgress national boundaries, water problems traditionally have been approached on a local—or occasionally regional—basis, and with comparatively little regard for long-term consequences. The classic example of an overworked, severely damaged river is the Rhine: no matter how much the Dutch may want a pollution-free flow, they receive waters already contaminated by upstream Switzerland, Germany, and France. Today, hydrologists realize that large-scale uses and diversions of water must be planned

Water leakage from California's All-American Canal is revealed dramatically by dark streaks of vigorous vegetation in this black and white version of an EROS color infrared photograph. (ROBERT W. PEASE, UNIVERSITY OF CALIF., RIVERSIDE)

and executed only after profound study of all relevant hydrologic data. According to hydrologist Charles J. Robinove, "water as a commodity for beneficial use can be managed in the largest sense only on a continental basis." Noting that "The logistic requirements for collection of [hydrologic] data are formidable and require the annual expenditure of millions of dollars in the United States alone," Robinove concluded that such data can be obtained "economically, repetitively, and with assurance of its beneficial use" by space-sensing techniques.

The mapping of snow and ice cover offers one of the greatest benefits emanating from a hydrologic satellite system. At any one time, snow covers from 30 to 50 percent of the Earth's land area, glaciers about 10 percent, and ice many of the inland waters, particularly in the Northern Hemisphere. The cost of obtaining the information needed by conventional aerial and surface means is great; indeed, it is often impossible or extremely difficult to obtain it at all. Yet, it is urgently required by hydrologists who must plan, manage, and operate irriga-

The Brahmaputra River Valley in the State of Assam, India, and the Himalaya Mountains, from Skylab 4. (NASA)

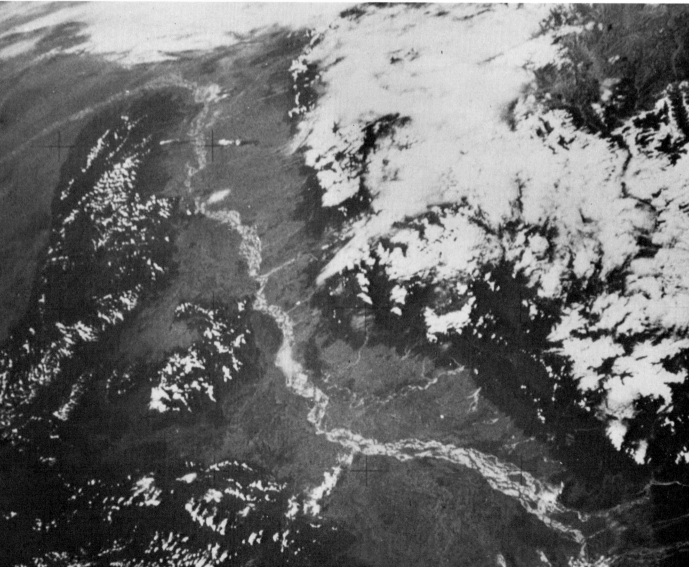

tion projects, hydroelectric plant schedules, regional and local water supplies, and flood warning systems. Paul McClain of the National Oceanic and Atmospheric Administration stated that imagery flowing from ERTS 1 provided "the best sequential, synoptic data on ice conditions ever taken by remote sensing for the purposes of ice dynamics studies." As far as snow investigations are concerned, NOAA's Donald R. Wiesnet was extremely pleased with ERTS 1's ability to locate the snow line on Mt. Rainier in the state of Washington. He observed that "the reso-

lution of the ERTS 1 MSS [multispectral scanner] is certainly sufficient for determining the snow-line elevation in mountainous terrain for all but the most stringent hydrological requirements. In 1974, George and Helen Kukla of the Lamont-Doherty Geological Observatory, Columbia University, reported that their studies of weekly NOAA satellite imagery maps of the northern hemisphere showed that snow and ice cover increased 12 percent in 1971 and has since remained at the new level. "The location and duration of snow and pack-ice fields constitute the most im-

China's silt-laden Yangtze and Lake Tungt'ing, lower left, from Gemini 5. (NASA)

portant seasonable variable in the earth's heat balance," they said, an opinion backed up by F. Kenneth Hare of the University of Toronto. He noted that this expansion of snow and ice preceded the coldest recorded Canadian winter. "In the winter of 1972, temperatures were below normal for the entire year over the entire country. This never happened before...I don't know what caused this, but the climate has in fact shifted into a more extreme phase." Elsewhere, scientists at NASA's Goddard Space Flight Center have shown that by using ERTS 1 photography of snow cover in the Himalayas, it is possible to determine the later runoff of the Indus River.

Because of the sparsity of ground stations, the difficulties of making surveys in remote and mountainous areas, and the expense of aerial overflights, synoptic, continuously upgraded data on snow and ice cover are seriously lacking, even in developed nations. Yet, it is estimated that if such data were available only in the western United States, the annual savings to water users would run from $10 million to $100 million a year. It is also known that many millions of dollars each year are lost due to damage from ice jams, and that many millions more are wasted by disruptions to navigation caused by unpredicted ice buildups. Avalanches cause much loss of life and property, while faulty programing of water resources due to limited knowledge of ice and snow conditions certainly results in further losses at the national, state, and local levels.

During the 1972-1973 winter, NASA and the U.S. Coast Guard undertook an experimental effort within the overall U.S. Winter Navigation Program to monitor the thickness, type, and distribution of ice in the Great Lakes preparatory to releasing timely information to ice breakers and cargo vessels attempting to find ice-free or, at least, thin-ice routes. As a result, ice data derived from ERTS 1 and the NOAA and Nimbus weather satellites is being used in concert with information obtained from airborne sensors in a continuing study of the feasibility of extending navigation of the Great Lakes and Saint Lawrence to a year-round basis.

In another, quite different kind of experiment, the ERTS 1 spacecraft repeatedly photographed the Bering glacier some 230 miles east of Anchorage, Alaska. This glacier is of the surge type that may remain stagnant for years and then suddenly advance at a surprising rate of speed. The western lobe of the Bering glacier, known as Steller lobe, dams up Berg Lake. It so happens that at the present time the main glacier is moving forward while the lobe is retreating. In so doing, it threatens to release catastrophically the dammed up lake water. Regular observations of the behavior of the Steller lobe would make it possible to issue warnings to people potentially endangered by a flood. Also spectacularly evident from ERTS 1 imagery is the extent of sediments discharged into the ocean by such Alaskan glaciers as Tokositna, Lacuna, Yentna, Kahiltna, and Ruth. One 1973 photo showed that the folded moraines associated with Yentna were displaced over a mile down the valley from positions observed in aerial photographs taken as recently as 1970.

Space surveillance is as useful for monitoring water pollution as for ice and snow. With the overview, it is possible not only to spot pollution downstream but to locate it at its source. In an era of rising threats to the environment, the satellite is a valu-

(Opposite page) The Great Lakes in summer, above, and winter, below. The upper infrared photo, made June 14, 1973 by the NOAA 2 weather satellite, portrays the spring warming pattern of surface waters, with darker tones representing higher temperatures. The lower, Tiros photo shows snow and ice cover on the lakes on February 20, 1970. (NASA)

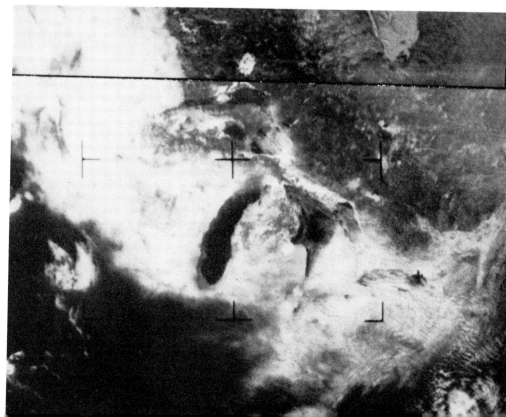

The Dead Sea, from an altitude of 568 miles, in a black and white version of an ERTS 1 composite color photograph. Jerusalem appears as darkish area at left center. (NASA)

Utah's Great Salt Lake, Salt Lake City, and adjacent areas, as seen from Skylab 3. (NASA)

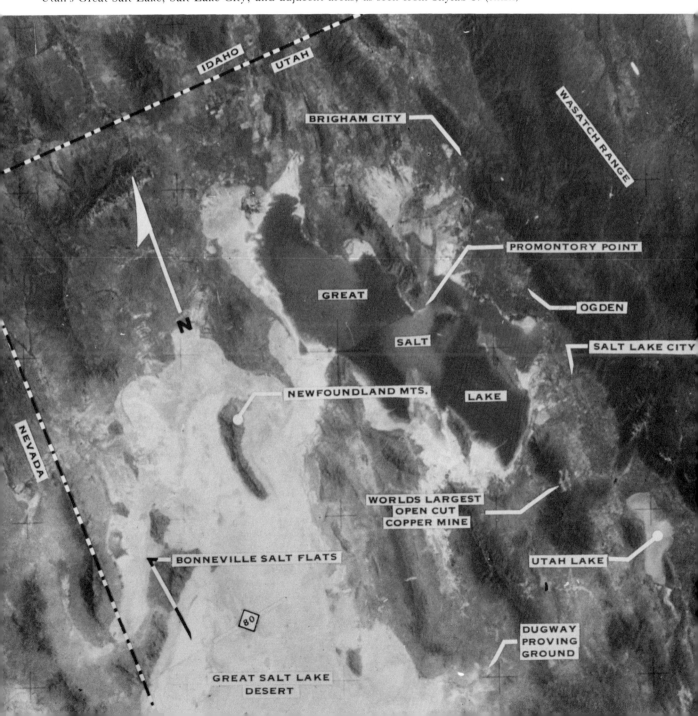

able new tool to the hydrologist seeking to minimize pollution of inland waterways. Even in such idyllic countries as Switzerland, the waters are becoming befouled, including famed lakes Lugano, Zürich, and Geneva. As materials and sewage from industries (particularly chemical plants) and homes pour into the lakes they help fertilize what are known as blooms of algae. These can be observed from above, helping provide quick identification to a trouble area. As the algae die, they fall to the floor of the lake and decompose. Oxygen progressively is removed from the water in algae-rich zones, killing off fish.

The capabilities of space surveillance in this regard are underscored by the introduction of a satellite photograph as back-up evidence in a lawsuit against a New York paper mill accused of polluting the Vermont side of Lake Champlain. According to former Vermont Attorney General James Jeffords, the introduction of an ERTS 1 photo into the legal process presents "evidentiary problems," in part because of its setting of a precedent and in part because no photographer can be brought into court to swear "that the picture is a true representation of the scene as it usually is." Vermont seeks a "cease and desist" order against the paper mill, owned by the International Paper Company, which discharges its 21-million-gallon a day effluent through a pipe that runs along the bottom of Lake Champlain almost to the Vermont border.

Just as hydrologic satellite sensors can detect pollution, so they can locate floods in their incipient states giving time for warnings to be sent out to the populace most likely to be affected. In the United States alone, floods are estimated to result in about $300 million in damage each year, with some years witnessing losses of greater than $1 billion. Hydrologic data relay satellites help reduce these losses by relaying information gathered by thousands of individual reporting and measuring elements of flood warning systems to control points from where the warnings are issued. After flooding is under way, hydrologic satellite observations are needed to show the limits of flood expanse so that appropriate action on the ground can be taken: damage assessed, rescue operations coordinated, and plans made to improve warning networks to reduce the probability of future catastrophes.

A comparison of ERTS 1 photographs taken of the Mississippi valley in October 1972 and again toward the end of March 1973 graphically shows the effects of the severe spring flooding that led to the declaration of several disaster areas. Surveys showed that many sandbars were either covered by water or washed away, while numerous oxbow lakes appeared greatly enlarged. Record-breaking spring rains delayed planting in many parts of the Mississippi valley, a fact also shown in ERTS 1 imagery.

Two long-range planners, A. H. Muir and R. A. Summers, have used hydroelectric power generation as a case study of a typical benefit accruing from the use of orbiting sensor platforms. They show that in

areas where large amounts of hydroelectric power are generated, more accurate advance information concerning

Flood waters at St. Louis: River levels are normal in the ERTS 1 photo at near right, made October 2, 1972. The Missouri joins the Mississippi at point A; the Illinois joins the Mississippi at point B. In the second photo, taken March 31, 1973, letter C shows flooded areas; the river then was at a stage of 38 feet above normal and still rising. It crested at St. Louis in late April at 43.31 feet, highest level in history. (NASA)

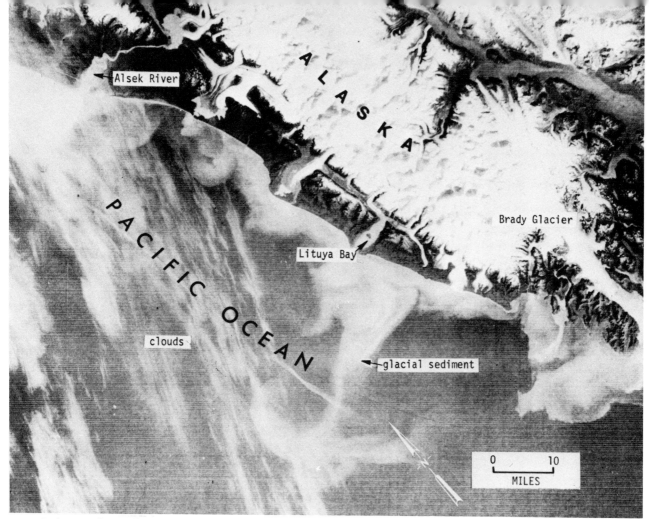

Sediment plumes from coastal glaciers of Alaska extend more than thirty miles into the Pacific in this ERTS 1 photograph above, providing the first visual evidence of the remarkable extent of glacial runoff. (NASA)

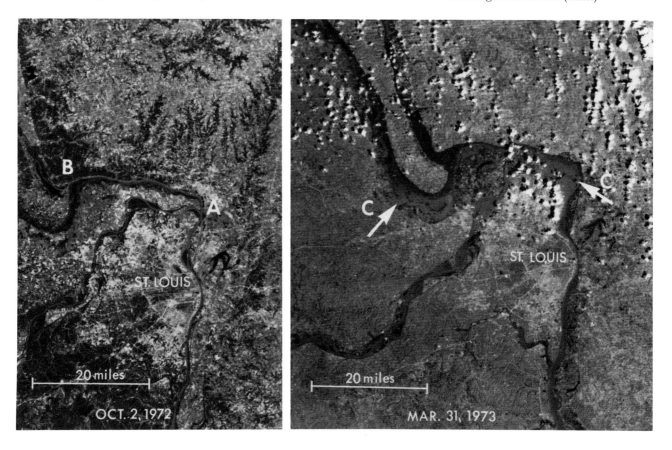

runoff from mountains and tributaries can be of tangible benefit to dam managers in planning optimum use of their water supply. Thus, the ability to detect conditions that, in turn, contribute to more accurate short-term and total seasonal flow forecasting of inflow into the reservoir would permit dam managers to plan use of the water supply in such a way as to reduce spillage and maximize power production.

They not only expect more accurate estimates to be made of year-round stream flow but a significant overall improvement in hydroelectric power generation management resulting in important economic benefits. Related studies show that an improvement of 25 percent in the accuracy of information on stream flow rates yields a 40 percent rise in the effectiveness of regulating large reservoirs.

In recent years, hydrologists have begun to look at rivers and streams in terms of the broad systems approach, taking into consideration a gamut of natural and man-induced factors. For example, agricultural and lumbering operations upstream may, in combination with a heavy rainfall, have detrimental effects on downstream reservoirs, navigation, and city water supplies. Ground and sporadic aerial observations alone are often not sufficient to supply the timely, synoptic, and sequential information necessary to understand all the variables that affect the total river system.

Geologists and hydrologists are beginning to rely on the overview to give them a full understanding of the geologic development of the river bed and flood plain, the fluvial processes operating in the recent past and today, and how the river is likely to act hours, days, and even weeks in the future. They point out that the view from orbit is particularly important in providing the overall identification of stream patterns—be they meandering, braided, or straight types—and predicting river flow and river slope from meandering characteristics. (Braided patterns normally involve steep slopes and/or high bed-load concentration, and occur in channels carrying material of the coarseness of sand or greater as the bed-load.) Periodic information of flow and slope are essential to planning the optimum use of water resources.

Experiments show that color infrared photography can enhance the characteristic patterns of braided streams, locating channels, identifying islands that are free of vegetation and those that are covered, and spotting submerged bars. Overview photography is also useful to hydrologists in studying broad fluvial and geomorphologic arrangements that influence river systems and hence the works of man dependent on their behavior—bridges, tunnels, riverfront structures, reservoirs, navigation, irrigation, etc.

M. M. Skinner of Colorado State University has been active for many years in identifying ways that space-borne sensors can assist hydrologists in studying river control, navigation, water resources management, pollution, and related subjects. The categories that he believes to be of particular importance to river engineers are:

*Sediment Transport Processes:* Color infrared photography helps spot the relative concentrations of suspended sediments in rivers. Thus, dark-colored areas indicate essentially clear water whereas

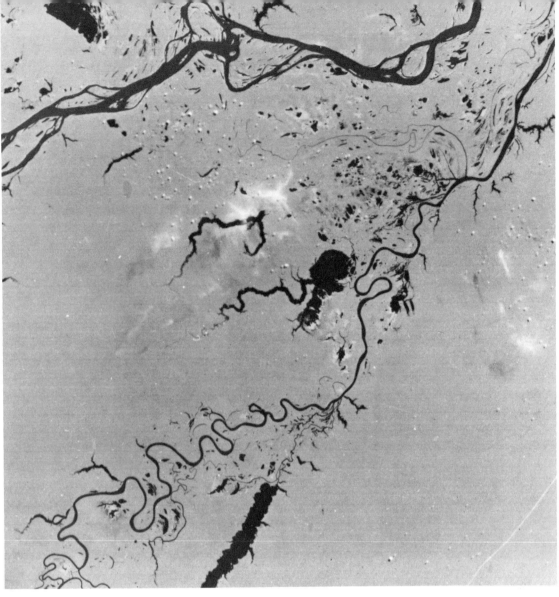

Courses of Amazon and Purus rivers
in Brazil are clarified by false-color
image, above, which highlights
water only. Right, the mouths of
the Indus River, emptying into the
Arabian Sea, lower left quadrant,
in a black and white version of a
color composite photo. (NASA)

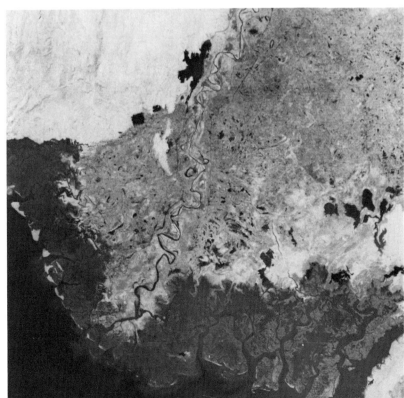

light bluish tones give evidence of sediments in suspension. Two ERTS 1 images of the area around Washington, D.C., taken on September 28 and October 11, 1972 provide an example of contrast. In the first picture, it was evident that the Potomac River's silt load extended below Colonial Beach. On the second image, the silt load was much more concentrated due to a heavy rain the day before; however, an incoming tide kept it from extending as far downstream. Being able to trace such sediments is of far more than academic interest. Along rivers, small and great alike, the overview readily shows where erosion is occurring and where the resulting sediments are going. Following sediment transport and dispersion paths indirectly helps hydrologists determine where flow separations take place in river waters.

*Flood Plain Vegetation Surveys:* In many arid and semiarid regions of the world, agricultural activities tend to cluster around fertile flood plains. But the very fact that these plains afford favorable conditions for crops also means that nonbeneficial plant life is likely to abound. This not only means that land is being lost to cultivation but that much ground water unnecessarily is being consumed. Moreover, Skinner directs attention toward ". . . the encroachment of salinity problems induced by the proximity of the water table, poor drainage characteristics, and the reuse of the water itself." He shows that color infrared photography helps evaluate plant species, identify certain kinds of plant stress, and reveal plant vigor—which "is often related to the proximity of the water table." Overview photography has become increasingly important in delineating areas in which ground water might be conserved by

eliminating nonbeneficial vegetation; naturally, the ecological effects of doing away with the "nonbeneficial" plant life must be considered carefully before the step is taken.

*Soil Classification:* The ability of soil and gravel characteristics to be interpreted on the basis of high-quality satellite imagery is another interesting benefit. In many parts of the world, flood plain gravel deposits are considered scarce, causing them to be placed in the valuable-mineral category.

*Drainage Pattern Net:* High quality space photography and imagery can readily supply information on the drainage pattern, of importance to understanding the discharge characteristics of streams and rivers.

*Water Resource Management:* The hope is that repetitive satellite monitoring of snow accumulations will result in such accurate predictions of melt and runoff that reservoir managers will not have to release too much water at one time because of the arrival of unexpectedly large amounts of melting snow-field waters. On the other side of the management problem, satellite observations of water use patterns and the appearance of crops should help reservoir managers in determining when to release water for irrigation purposes; such data would be especially valuable in time of drought, when water is in short supply.

*Channel Changes:* "A river tends to meander and wherever man has affected this normal meandering pattern he can expect considerable difficulty in maintenance of the channel." Thus, Skinner warns against tampering with nature's ways without first assessing the consequences, and reports that an inspection of space

(*continued on page* 97)

A beautiful view of the Earth's sphere extending from the Mediterranean Sea area to the Antarctica south polar ice cap, as photographed from the Apollo 17 spacecraft. The southern hemisphere has heavy cloud cover, but almost the entire coastlines of the continent of Africa and the Arabian Peninsula are clearly delineated. (NASA)

Earth's terminator is enhanced in this Skylab photograph by clouds casting shadows from the dayside back into the portion of the globe that is just emerging from the night. (NASA)

Sunrise over the Mascarene Islands, several hundred miles east of Madagascar Island taken by Gemini 6. Cumulus clouds are aligned east-west over Mauritius, center; to island's south is an excellent example of an open convective cloud cell. Réunion Island is in upper center. (NASA)

Sun blazes on Socotra Island in the Indian Ocean and the channel between it and nearby smaller islands, called The Brothers. Left of the sun glint in this Apollo 7 view, a slicklike eddy is outlined darkly. The fine white horizontal line below Socotra may be the result of waves rolling over an undersea shelf. (NASA)

Remains of a large thunderstorm cell over the Amazon, photographed from Apollo 9. Symmetry of cloud shield shows it is virtually stationary; if it were moving, the shield would be stretched downwind in an anvil shape. (NASA)

Well-defined pattern of fracture lines in this Gemini 4 photograph of southwestern Saudi Arabia and Yemen suggests the presence of underlying domal structures that might contain oil. (NASA)

Irrigated farmlands such as these near the mouth of the Colorado River (red checkerboard pattern) can be spotted through the use of infrared photography. This particular shot was taken during the Apollo 9 mission. Light colors of silt bands in waters of the Gulf of California reflect currents and depths. (NASA)

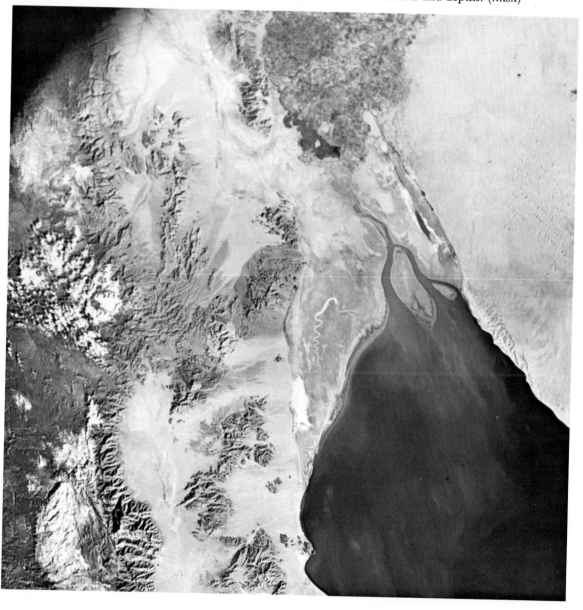

Infrared photography helps detect agricultural diseases and pest attacks before they can be observed from the ground. Here, insect-infected trees in Oregon appear blue-green, while healthy trees appear red or pink. (NASA)

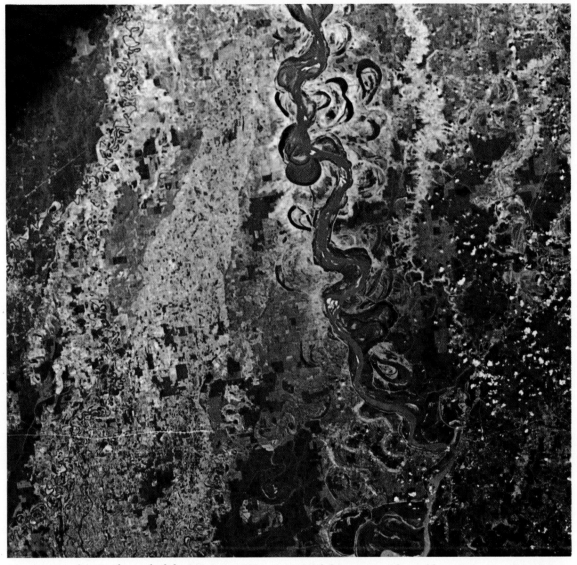

Movements of main channel of the Mississippi River near Vicksburg are evidenced by numerous oxbow lakes and meander scars in this Apollo 9 infrared photograph. Aerial surveillance of changes like these may lead to improved flood control programs. (NASA)

Los Angeles County and Kern County, California, in a Skylab color infrared photograph. San Andreas and San Garlock Faults intersect at lower left center. San Gabriel Mountains extend across bottom; Isabella Reservoir is at upper left. (NASA)

Lake Powell area on the Colorado River, with the Grand Canyon to the north, viewed from Skylab. Darkest area is Kaibab Plateau. Black Mesa is the dark area in southernmost part of photograph. (NASA)

The first color composite photograph returned by ERTS 1, this view of California-Nevada was made on July 25, 1972. Green, red, and infrared are combined to reveal details not visible to the naked eye. Here, healthy vegetation on the Sierra Mountains makes them appear bright red. Lake Tahoe is in upper right corner and Stockton, California, in the lower left. (NASA)

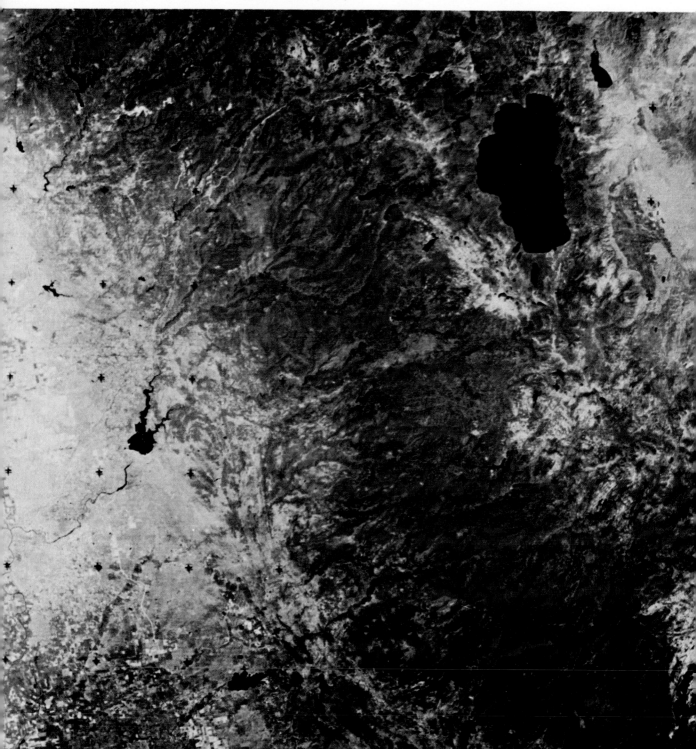

A spectacular ERTS 1 photograph of the Grand Teton National Park area. The Tetons are in upper center, with the Wyoming Mountain Range extending from the base of the Tetons in the center to the lower right corner. The Snake River enters at upper left, runs southeast to Palisades Reservoir, center, then curves northward, just to the east of the Tetons, and flows through Jackson Hole, seen here as a bluish-gray patch to the right of the Tetons. (NASA)

Orbital photos of key demographic areas such as this Skylab view of the Chicago metropolis, including Aurora and Joliet, Illinois, and Hammond and Gary, Indiana, are of great value in ecological and cultural studies.

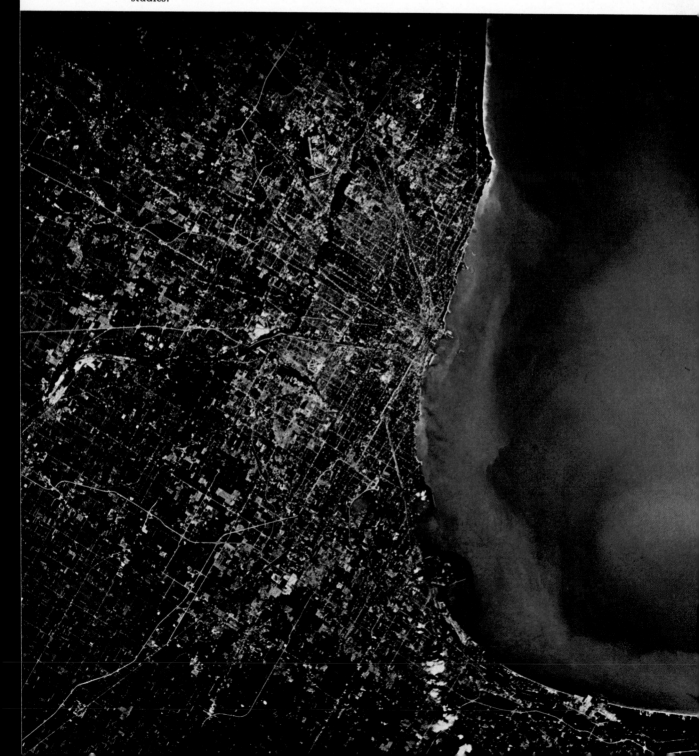

This ERTS 1 land use photo gives a wealth of information about an area containing some 10 percent of the U.S. population. Dark gray tones indicate areas of greatest population density and/or commercial development. Metropolitan centers include New York City, (1), suburban Long Island (2), Trenton (3), Philadelphia (4), Allentown/Bethlehem (5), and Reading (6). Major transportation arteries, such as New Jersey Turnpike (7) also can be seen. Forested Appalachian Mountains appear bright red. Variations in red shades indicate grasslands (light), agricultural areas (darker), pine forests (very dark) and wetlands (darkest). White spots indicate either construction activities, barren/dry field, or sandy beaches. (NASA)

Marring the picture: Air laden with pollution, extending over an enormous area of the United States and Canada, and stretching eastward into the Atlantic, as photographed by Apollo 11 astronauts at a distance of 4,000 miles from the planet. (NASA)

photographs of the lower Mississippi river system quickly revealed why difficulties in navigation and flood control were arising. Such photographs not only show the whole pattern of a scene at a single time, but regular, repetitive monitoring permits river engineers to be constantly alert to changes.

*General Knowledge of River Systems:* Most important, the orbital overview allows man for the first time to see complete river systems in operation, whereas before he had to piece together hundreds of thousands of individual observations, taken at different times and under different seasonal conditions.

The overall economic advantage of hydrological satellites can be assessed only roughly because of the virtual impossibility of gathering the information needed to make good estimates. However, it is known that the U.S. spends about $100 million a year on water research, and that the quality of this work would be vastly improved were regular hydrologic satellite data available. Studies have shown that on a 31,000-acre project in Idaho, farmers realized $317,000 in savings or increased income because of accurate forecasts made of an exceptionally low water flow one year—they either reduced acreage or changed their cropping patterns on the basis of having the timely information. In Canada, a mere 1 percent improvement in forecasting water flow rates from April to August resulted in $1 million in increased power revenues for one modest project. And some $600,000 was saved by anticipating potential flood damage to Phoenix, Arizona: reservoir engineers released large quantities of water on the basis of reliable forecasts made available to them. These sort of examples can be multiplied across the United States and around the world.

Research activities taking advantage of the overview and remote sensing instrumentation are almost endless. In Florida, a study is being made to determine the interrelations between the lakes of the west central part of the state and their effects on the hydrologic regimen of the area. Another study, in North Carolina, has to do with the cooling reservoir of a large steam-electric generating plant, while at the Yellowstone National Park, remote-sensing is employed in the search for potable ground water. According to Edward R. Cox, "The use of remote-sensing techniques, particularly infrared imagery, should be a valuable tool in a reconnaissance of ground-water conditions near thermal areas, and it could, in places, eliminate the need for or reduce the number of test holes to be constructed." The drilling of test holes is a time-consuming and costly operation.

Both ERTS and Skylab photography has located fresh water in areas that need it badly. James V. A. Trumball of the U.S. Geological Survey reported that Skylab's sensors identified specific new fresh water sources and green forage in Puerto Rico, while in Arizona a team including California Institute of Technology's Eugene Sloemaker and USGS's Donald Elston and Ivo Lucchitta geologically scanned for water in the Verde Valley, in the Shivwits Plateau northeast of Lake Mead, and in the Coconino Plateau between Grand Canyon and Highway 66. When a water hole was drilled 10 miles south of the Grand Canyon, an aquifer was struck at some 40 foot depth, an amazing result. In related studies, the orbital overview also showed that old volcanic lava beds on the Colorado Plateau are so fractured that they may provide direct access to subsurface water-bearing rocks. This fracture system approaches to within 10 miles of

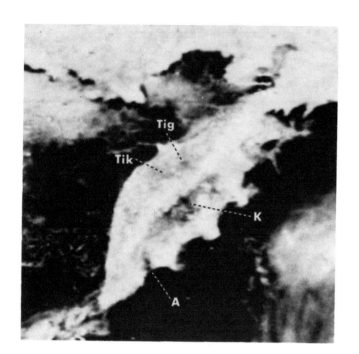
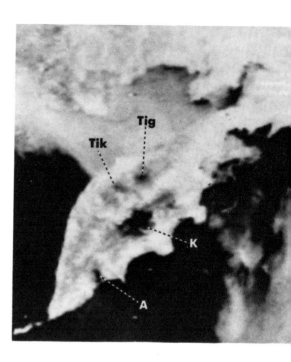

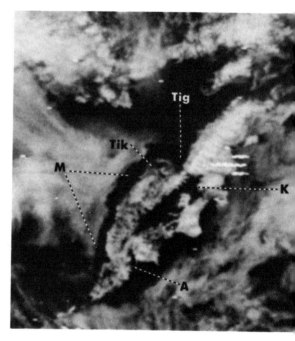

Surveillance of melting snows on Kamchatka Peninsula, U.S.S.R., by Nimbus 4. Top left, April 22, 1970, melt has begun in Kamchatka (K), Avacha (A), Tikhaya (Tik) and Tigil (Tig) river valleys. Top right, five days later, dark areas have increased, indicating further snow melt. Bottom right, a month later, May 27, 1970, snow has melted in all river valleys and in marshland (M). On map, lower left, marshland and valleys are indicated in black, mountains in white. (NASA/GODDARD)

The Mississippi Valley near Vicksburg, in a black and white version of an Apollo 9 color infrared photograph, taken from about 120 miles above the surface. Repetitive photography of this sort helps engineers keep abreast of channel changes. For comparison of the black and white rendition with the original color infrared photograph, see page 89. (NASA)

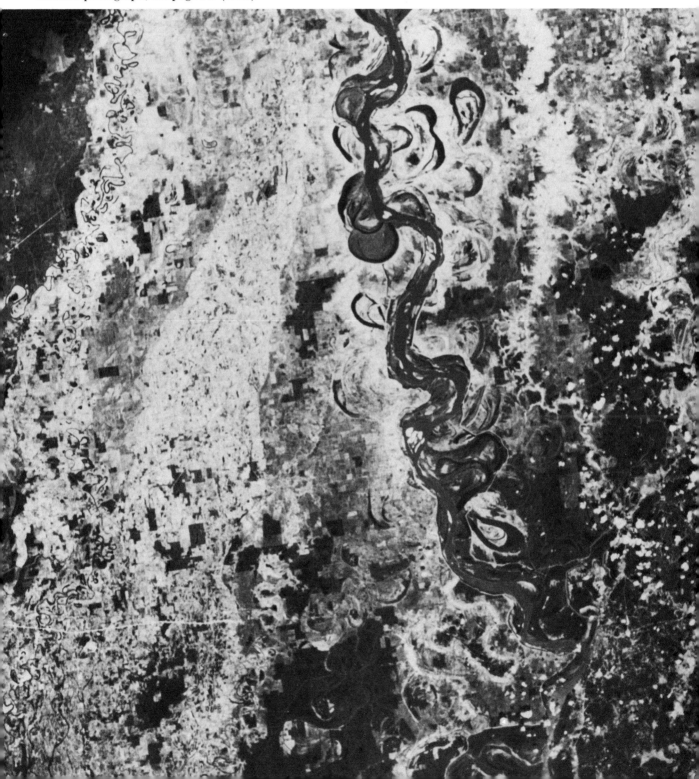

Flagstaff, a city looking for a reliable water supply.

An example of how effective sensing in the near infrared can be is shown by two ERTS 1 photographs of lakes near Susanville in California. When ERTS 1 first photographed the area on July 25, 1972, the lakes were found to be well filled with water and thus stood out in the image in sharp contrast with the surrounding land. When the lakes were rephotographed on August 13, one turned up completely dry and the other with but half its previous content. Both views were different from conditions exhibited on the most recent map of the area, which was prepared in 1962. The water was found to have been drawn off for irrigation purposes.

A larger scale project involves the direct application of space photography to a fascinatingly complex hydrological problem in the American Southwest. For many years, the U.S. Geological Survey, the Bureau of Reclamation, the Army Corps of Engineers, the Texas Water Development Board, and a citizen's group known as Water, Inc., have been trying to solve the problem of providing adequate water to farmers in the Southern High Plains of West Texas and Eastern New Mexico. Today, about 6.3 million acres are irrigated with ground water coming from the Ogallala Formation aquifer, a stratum of wet sand and gravel varying from fifty to three hundred feet below the ground. This aquifer is recharged by rainwater and seepage from some 20,000 ephemeral lakes known as playas. Geologists know that increased irrigation since the 1940s has greatly depleted the water available in the Ogallala Formation, not only eliminating the possibility of expanded agriculture in the area but actually reducing it. Complicating the situation is the fact

that the hydrologic cycle of the playa lakes is little known. Hydrologists urgently needed to know how much water was being received by the playas as a result of runoff from precipitation, how much was lost by evaporation, how long surface waters remained stored in them, and how often and to what extent they recharged the Ogallala Formation.

The Geological Survey has attempted for some years to measure precipitation, evaporation, surface runoff, and water storage in the playas over a 35,000-square-mile experimental area by installing an instrumentation network. Existing maps are of little help since they are not synoptic, and the playas are continually changing anyway. Hydrologists know that it is extremely tedious and expensive to evaluate synoptically the playa lake system from the ground, and that repetitive high-altitude photographs are also very costly not only to make but later to rectify and analyze. The solution of their dilemma had to await the advent of orbital photography.

At 10:31 A.M. on March 12, 1969, the Apollo 9 astronauts photographed the area from an altitude of 117 miles. Light areas indicative of dryness (high reflectance) showed up; they represented cultivated fields that were then dormant, exposed sand, winter wheat, and urban centers. Newly cultivated fields, range lands, and scrub brush appeared darker in color, and water surfaces and wet soil were very dark in contrast to the dry soils. The dark centers of many playas—identified by light, near-circular rings—were usually moist soils, although some depressions contained water.

The existence of this one space photo of the entire area demonstrated to hydrologists the immense value of the overview—

100

and of the importance of repetitive coverage. ERTS 1 and Skylab later assisted them in working out the High Plains of Texas hydrologic cycle by observing distribution of the playa lakes and then recording which of them contain water at a given time; the extent of water coverage including depth, the rate of change of water level and storage over the entire area; the relative rate of storage between individual playas, and their response to storms. One cloud-free ERTS 1 image revealed a total of 6,631 playas in the 13,225-square-mile area, no two of which were identical as far as bottom conditions were concerned. This reflected not only how much water was being recharged but how much was being evaporated. (Playa bottoms, incidentally, change constantly due to the deposition of silt coming from eroded agricultural lands.)

The distribution of surface water in the Southern High Plains obviously is dynamic. The causes of differences in storage and distribution of playa water range from local rainfall patterns and continuous evaporation to the steady percolation of water through the ground and its use for irrigation. In the words of John M. DeNoyer, director of the EROS program:

The satellite data make it possible to repetitively observe the rapidly changing water distribution throughout the year and to determine the number and sizes of lakes that are present. There is no other practical way to acquire these data. Data of this kind are essential for improving our ability to manage the surface water in terms of its utilization and possibly to plan for artificially recharging the subsurface aquifers in certain areas or for the importation of water where needed.

West Texas near Midland and Odessa, with the Concho River drainage at lower left. Photo made during Gemini 4 mission. (NASA)

The High Plains situation is a good example of how agricultural activities can strain the capacity of nature to sustain them. While the space overview certainly will help to alleviate the immediate problem, in the long run men will be forced to accommodate themselves to resource limitations and ecological realities. Planet Earth is not an endless bank of riches, but managing finite water resources from orbit is a giant step forward.

101

# 5

# Agriculture and Forestry

CONRAD: *Got a good cornfield?*

KERWIN: *[Laughing] Yes, I'm trying to find a uniform crop line here. You can really see the circular areas where they irrigated that walkaround stuff, you know. There's one there, and I'll try to—OK, tracking is good, the Holt County, Nebraska, field there. The field is slightly larger than the reticle in the cross hairs.*

WEITZ: *Mark the time to uniform crop-line areas.*

CONRAD: *Okay.*

WEITZ: *They ought to be big in Nebraska this year. See, what we do is we apply this in a practical manner. We see how the corn crop's going to go, and from that you find how the present crop is going to be.*

CONRAD: *Right.*

—DIALOGUE ABOARD SKYLAB 2, JUNE 9, 1973

The land area of planet Earth totals about 33.92 billion acres. The Food and Agriculture Organization of the United Nations estimates that approximately one-third of the area—more than 10 billion acres—is forest land. Another third, comprised of 3.75 billion acres of land that is either arable or under tree crops plus 6.5 billion acres of meadow and pastureland, is devoted to some form of agriculture. The remaining 13.42 billion acres are either wasteland or being used for cities, railways, airports, military posts, highways, and other works of man.

As the number of people in the world increases so, obviously, does the strain upon the land to feed them. Thus, while world consumption increases by some 2½ percent each year, global grain reserves instead of keeping pace have actually decreased during the past decade. Mankind is faced with a loss or degradation of vi-

Diseased plants appear darker than healthy ones in this aerial infrared photograph of a potato field into which blight was introduced for test purposes. Diseased areas were apparent in photographs several days before changes were noted on the ground. (NASA)

tally needed land, especially in underdeveloped countries where excessive tilling and primitive agricultural practices have resulted in much topsoil erosion. And erosion, in turn, exacts heavy tolls: prices of crops go up (between July 1972 and July 1973, corn and wheat rose 136 percent, wool 137 percent, rubber 153 percent, and cocoa 229 percent) while in 1974 sugar prices increased nearly threefold at the retail level and more than fivefold at the raw sugar level; the amount of time needed to produce them increases; profits of farmers decline, and rivers, lakes, and harbors become clogged with silt.

Certainly the most stricken area in the world today, caused in part by the worst drought in the recorded history of Africa, comprises the West African nations extending from Mauritania to Chad. When the rains finally returned in 1973, many farmers had not planted at all, while others had consumed their seeds just to survive. In September 1973, Skylab mission 3 astronauts Alan L. Bean, Owen K. Garriott, and Jack R. Lousma carefully photographed the entire area with the hope that later analyses of water and mineral resources and vegetation patterns would contribute to an improvement in the disastrous situation.

The forests also suffer as demand grows for wood and wood products while the expanding population at the same time seeks to convert forest land to agricultural uses. The increasing pressure upon the forests has grave implications, since forests perform a number of important roles in the global ecological system by preventing soil erosion, harboring wildlife, increasing air humidity, reducing wind velocities, moderating temperatures, and providing watershed protection.

The actual and potential uses of the satellite overview to the development and management of agricultural and forestry resources are varied and require the application of multispectral, infrared, and other sensing techniques to identify various crops, trees, and range lands; to assess their maturity and health, and to perform many other functions ranging from preparing agricultural censuses to monitoring conservation programs. By conducting orbital surveys, the predicted productivity of the land can be more accurately estimated early in the crop cycle, helping maintain the supply of essential foods and fibers in balance with demand. With worldwide crop information available, accurate predictions would become possible nearly everywhere, leading to early assessments of the needs of deficit nations, quantities of commodities available for export by surplus nations, and international seasonal trade between Northern and Southern Hemispheres.

It also should be possible to make substantial savings in the costs of producing crops by taking advantage of the more complete, more timely, and regular information afforded by the satellite overview. For example, a University of Michigan survey suggests that agricultural benefits from orbital surveys of snows would run from $1 million to $5 million annually, and that in years of particularly heavy or light accumulation, accurate knowledge of predicted melt destined for surface water irrigation would yield "much larger benefits." Soil moisture measurements are also important. In the case of cotton (whose average yearly production in the U.S. during the 1960 decade was nearly 14 million bales), about 40 percent of the crop was grown with the aid of irrigation.

Land use studies from space: California's Imperial Valley, in a Gemini 5 photo, (above), made August 21, 1965, and in an infrared Apollo 9 photograph, (right), March 12, 1969. Analysis of colors and shapes over a period of time reveals the kinds of crops planted and the extent of the plantings. (NASA)

The Department of Agriculture estimates that if a 10 percent improvement in information on moisture conditions in the land had been realized from orbital surveys, the optimum dates for releasing irrigation waters could have been selected and the amounts of water injected into the cotton-growing area more accurately established. This would have resulted in an estimated annual savings of about $100 million.

Although efforts are continuously being made to improve the situation, no up-to-date and complete global agricultural inventory exists. Most agricultural maps, and the black-and-white aerial photographs on which they are in part based, are either obsolete, incomplete, or—most likely—both. Yet such inventories are essential for national and worldwide assessments of crop type, crop vigor, the nature and arrival of crop-damaging agents, crop yield per acre by type, crop acreage by type, and total crop yield.

Robert N. Colwell of the University of California at Berkeley has established tables depicting the type of vegetation resource data desired (broken down by agricultural crops, timber stands, rangeland forage and brushland—mainly shrubs) and the major federal, state, county, and private agencies that require them. Having done this, he spells out the frequency with which the information is needed. As examples, he suggests that every 10 to 20 minutes it would be desirable to observe waterline advances in cropland areas during floods, the onset of locust swarms, and the start of rangeland and brushland fires. At the 10- to 20-hour interval, typical tasks would be to monitor the U.S. wheat belt for outbreaks of black stem rust due to spore showers as well as map the perimeters of rangeland and brush-field fires and flood activity. Then,

between 10 and 20 days, one could map the progress of selected crops leading to accurate estimates on their vigor and when harvesting should begin. On ranges, this same interval would help cattle ranchers determine the readiness of the range for grazing, while out in the brush it would be possible to observe the ". . . times of flowering and pollen production in relation to the bee industry and to hay fever problems." He has also provided examples of the kind of information that agriculturists should have at regular 10- to 20-month, 10- to 20-year, and 20- to 100-year intervals.

As the possibilities inherent in remote sensing become more apparent to agriculturists, they naturally seek all possible opportunities to conduct experiments that will help translate theory into tangible benefits. Based on the use of Apollo 9 and high-altitude (60,000–70,000 feet) aircraft photography of an agricultural test site in Maricopa County, Arizona, William C. Draeger and his associates at the University of California at Berkeley achieved results that were encouraging in two ways: "(1) the questions posed initially are being answered, i.e., the very practical problems of an operational survey are being faced and solutions are being found, and (2) it would seem that a fully operational agricultural inventory using space photography is not beyond the scope of present technology." In other tests, using color infrared imagery to sense "disturbed" vegetation on Santa Cruz Island off the southern Californian coast, a team of geographers reported that they were "extremely encouraged by the results obtained from the imagery," which permitted them to make a "searching analysis" of "six plant communities that not only exhibit much internal variance but

106

have been grossly disturbed by intensive sheep grazing for more than one hundred years."

Following its launch in July 1972, the ERTS 1 satellite provided even more promising results. For instance, just two days after receiving a computer-compatible tape of a multispectral scanner image of parts of Texas and Oklahoma, Purdue University investigators were able to make this breakdown: range and pastureland, 4.1 million acres; crop land, 2.7 million acres; forests, 1.5 million acres; and inland waters, 190,000 acres. To Purdue's David A. Landgrebe, there "appears to be no question about the usefulness of this ERTS-provided capability." Meanwhile, Colwell, at the University of California, reported that the quality of agricultural and forestry activities also showed up rapidly and effectively on ERTS 1 photography, permitting one "to determine whether a pioneer who tried to establish agriculture in a previously uncultivated area made a go of it or had to fold up his tent and leave." He noted, too, that multispectral scanner images of the northern part of California were obtained in less than five minutes from ERTS 1, whereas an airplane would have required at least 250 *hours* flying time to survey the same area. Both ERTS and Skylab imagery has shown that forest surveys can be made at cost savings of almost 100 to 1 with 90 percent accuracy. One Skylab experimenter, Andrew Bensen of the University of California, predicted that such results will surely "decrease the number of visits to the woods."

The U.S. Department of Agriculture also conducted a number of experiments based on spacecraft imagery, including: 1) spectral reflection studies of vegetation, soil and water; (2) spectral techniques to record the defoliation of gypsy moths, and (3) the utilization of space techniques to determine the extent of soil erosion caused by the winds in the High Plains of Texas. Tests in the San Joaquin Valley in California showed that crops could be identified in terms of color tones. Thus, dark red meant rice; white, barley; amber, safflower, and pink, sugar beets or alfalfa. Burnt stubble showed up black; water, dark blue; bare soil, blue gray, and mountain forests, red.

Many additional agricultural observations were conducted during the Skylab missions beginning in May 1973. Data derived from the Earth Resources Experiment Package aboard Skylab has helped determine the need for crop irrigation, the distribution of saline soil, temperature distribution associated with freezes, disturbances caused by mountain pine beetles, and the early detection of infestation in crops.

Remote sensing of agricultural phenomena is being studied in other countries than the United States, including India, Argentina, Mexico, and Brazil. In the latter nation, for example, as early as 1968 the Commissão Nacional de Atividades Espaciais (later renamed Instituto de Pesquisas Espaciais) commenced a collaboration with NASA that resulted first in the training of Brazilian scientists in the United States and then in remote sensing from aircraft to help set the stage for the later application of the orbital overview to national needs. Scientists within the Office of Research and Experimentation of the Ministry of Agriculture are studying the viability of substituting conventional survey methods by satellite data at a number of sites, including No. 801 in Campinas. Investigations at that site are aimed at resolving and developing identi-

107

Irrigated, cultivated land is clearly visible from orbit in this Apollo 7 view of the Sudan, below Khartoum. (NASA)

Shapes and colors of Texas fields reveal winter wheat planting in this Apollo 6 photograph. (NASA)

Crop identification in Texas by remote sensing, using Ektachrome infrared film. Crops are identified by their signatures as: (1) mature lettuce, (1a) young lettuce, (2) cabbage, (3) oats, (4) peppers, (5) onions, (6) carrots, (7) parsley, (8) mature broccoli, (8a) young broccoli, and (9) bare soil. (NASA)

fication keys for crops such as widely grown coffee, citrus, and sugarcane; defining various weed species; designating the major stages of sugarcane growth; locating noncitrus orchards; and classifying wild trees and shrubs in terms of their being mature or immature.

Probably the most comprehensive agricultural research experiment involving remote sensing was the joint NASA–U.S. Department of Agriculture Corn Blight Watch conducted during the 1971 corn growing season in the states of Ohio, Illinois, Indiana, Missouri, Iowa, Minnesota, and Nebraska. (Independently,

the Michigan State University concurrently monitored a number of sites within the state of Michigan.) Following a loss of 700 million bushels (about 15 percent of the U.S. corn crop) in 1970 due to the blight (caused by the fungus *Helminthosporium maydis*) it was decided to establish the 1971 experiment using 1,806 corn fields located in 210 different sites each 8 miles long by 1 mile wide. RB-57F aircraft (this was a pre-ERTS experiment), flown for NASA's Earth Observations Aircraft Program office at the Manned Spacecraft Center by the U.S. Air Force Air Weather Service, carried the

110

cameras that photographed some 45,000 square miles of corn belt area in natural and infrared color with the aim of monitoring the development and spread of the blight during the entire growing season. Once the raw information had been gathered it was made available for evaluation by the Laboratory for Applications of Remote Sensing at Purdue University, the Institute for Science and Technology at the University of Michigan, the various USDA agencies, and NASA. Scientists were interested in determining how levels of blight infection could be assessed by remote sensing, what the impact of the blight (as well as other foliar stresses) was likely to be, and how techniques evolved during this experiment could be applied to other situations. The experiment showed that skilled analysts could take the infrared photographs and, with computer assistance, classify an individual stand of corn as healthy or as suffering from four degrees of blight: very mild, mild, moderately severe, and severe. Based on early warnings of the direction of spread of the disease, the farmers can take a number of steps, including spraying, making an early harvest, cutting the corn for silage, or even destroying the blighted area to help prevent further spread to nearby fields.

Satellites are equally useful for making regular inventories of forest acreage and detecting losses through fire and disease.

Forest fires not only destroy valuable commercial property but kill wildlife, remove vast areas from recreation, cause the soil to deteriorate, invite the invasion of insects and diseases, and increase the runoff of surface waters creating erosion. The immediate detection from orbit of forest fires and the transmission of data on actual and predicted weather conditions (including surface and high-altitude

winds) assist forest fighters to plan their suppression maneuvers. Fires can be extremely expensive. In 1964 a fire near Santa Barbara, California, burned some 67,000 acres of property and forest lands at a loss of more than $20 million, including $2.5 million just to suppress it. Fortunately, airborne infrared spotting techniques were brought to bear to identify the main fire center through the smoke, leading to the effective suppression. It was later estimated that the losses would have been $9 million greater if infrared had not been pressed into service at a critical time.

Even more serious than the danger of fire is that of insects and diseases—about three and a half times greater in the United States, according to the U.S. Forest Service. Bark beetles, spruce budworms, and the chestnut blight have taken a huge toll from the nearly 509 million acres of commercial forest in the United States. Synoptic orbital surveys by ERTS 1 have helped detect the advent of insect and disease damage at an early stage of development. Similarly, such surveys reveal nutrient deficiencies, areas of indiscriminate logging, and other data important to forest management.

An interesting case study of the application of space photography to forest resources was undertaken by the Pacific Southwest Forest and Range Experiment Station of the Department of Agriculture's Forest Service shortly after the Apollo 9 Earth orbital flight. Selecting an area of 10 million acres in Arkansas, Georgia, Louisiana, and Mississippi to make a pilot timber inventory, the foresters reported that the best results of the experiment were registered on a 5-million acre track in the Mississippi Valley "where a 58 percent reduction in sampling error was

111

Healthy trees in this citrus orchard in Texas appear lighter than those suffering from brown soft scale and blackfly infestation. The tonal difference is due to the healthy trees showing higher reflectance than diseased trees in the infrared light in which this photograph was taken. (NASA)

attributed to the information obtained from the Apollo 9 photography."

Remote sensing from space cannot provide all the answers to all the questions foresters ask; indeed, even if it could, there are simply not enough computers available to store all the inventory information that would be produced. Hence, forest managers resort to sample estimates which they then extrapolate to larger forest areas. Among the information gathered are timber volume, growth, and harvest statistics. Traditional ground and aerial surveys provide inputs on an area-by-area basis, with the result that when an inventory is made available to forest managers and planners many of the elements that make it up are outdated. Moreover, the very pace at which forests are being utilized makes it imperative that current information on them is always available.

The only way to ease this situation, in the opinion of foresters Philip G. Langley, Robert C. Aldrich, and Robert C. Heller, ". . . is a resource inventory and information system based on space-age technology." They feel that the system ". . . should be able to provide up-to-date estimates at any time quickly and efficiently on a national as well as local basis."

Rangeland resources are also being studied from space, both those devoted to commercial operations and those relatively undisturbed by man and his animals. The U.S. Department of Agriculture estimates there to be approximately 112 million head of cattle in the U.S., of which more than 35 million are located on ranges. The USDA also estimates that with the improved understanding of range conditions obtainable through remote sensing, it should be possible to

Conventional photograph of forest fire at Nuns Canyon in California's Sonoma district, left, reveals little more than smoke, while infrared image of the same scene, right pierces the smoke to pinpoint the location of the fire itself. Infrared sensor imagery also can inventory the extent of the burned area while smoke still lingers overhead. (U.S. DEPT. OF AGRICULTURE/U.S. FORESTRY SERVICE)

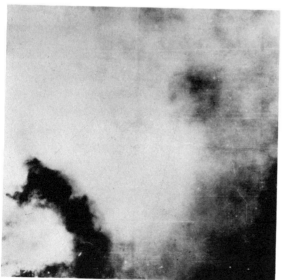

113

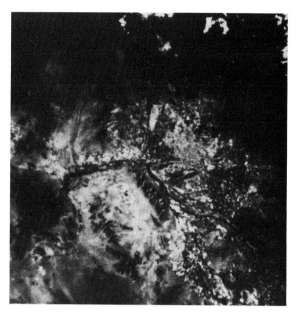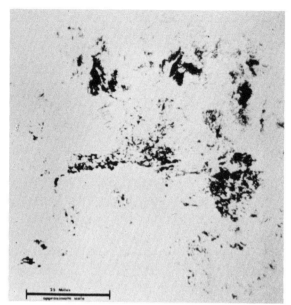

Three views of Phoenix, Arizona: Above left, an infrared Apollo 9 photograph, and right, automated mapping of the area's vegetation by extracting data from the Apollo photo. Next page, a near vertical Skylab photograph in natural light of the same metropolitan area. (NASA AND U.S. DEPT. OF INTERIOR/GEOLOGICAL SURVEY)

increase the stocking of rangeland from the usual 85 percent of carrying capacity to 95 percent—an increase of more than 10 percent. Resource inventory analysis and the monitoring of response to changing conditions would help managers assess areas of overgrazing, of insufficient nutrient, of accumulations of weed and brush, etc. Charles E. Poulton of the Oregon State University in Corvallis and his associates, enthusiastic supporters of remote sensing of rangelands, write that

Space and aerial photography, as well as other kinds of remotely procured imagery, have a unique characteristic which we in vegetational resource management have only begun to ex-

ploit. It captures and preserves a scene at a moment in time by recording the facts about the landscape which no other data gathering system can provide. These facts include such things as distribution, kinds, and amounts of vegetation. Such data are needed for decisions in the use of individual resources and for the integrated management of a number of interrelated resource uses.

Although the United Nations' Food and Agriculture Organization guardedly expressed optimism on world agricultural prospects at the beginning of the 1970 decade, by September 1973 the situation had so deteriorated that it had to call an

114

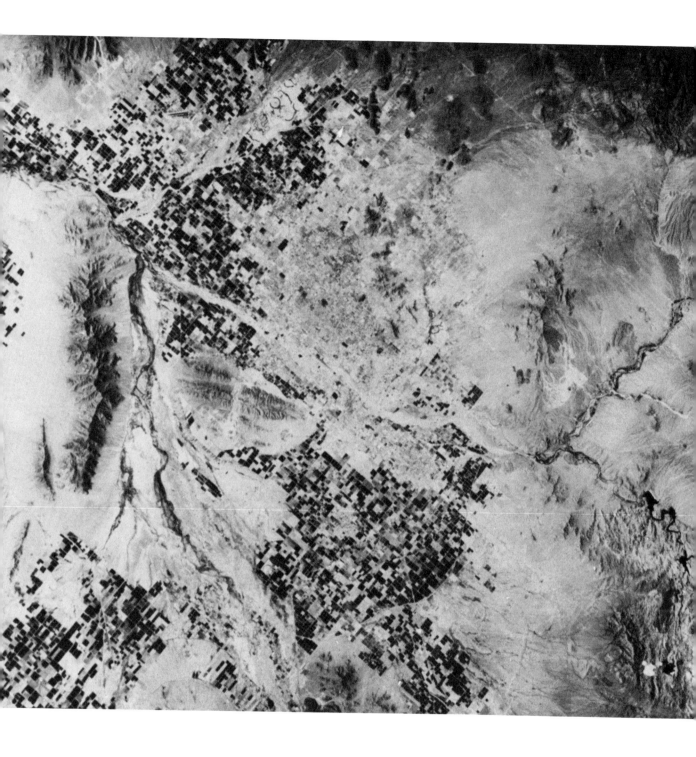

115

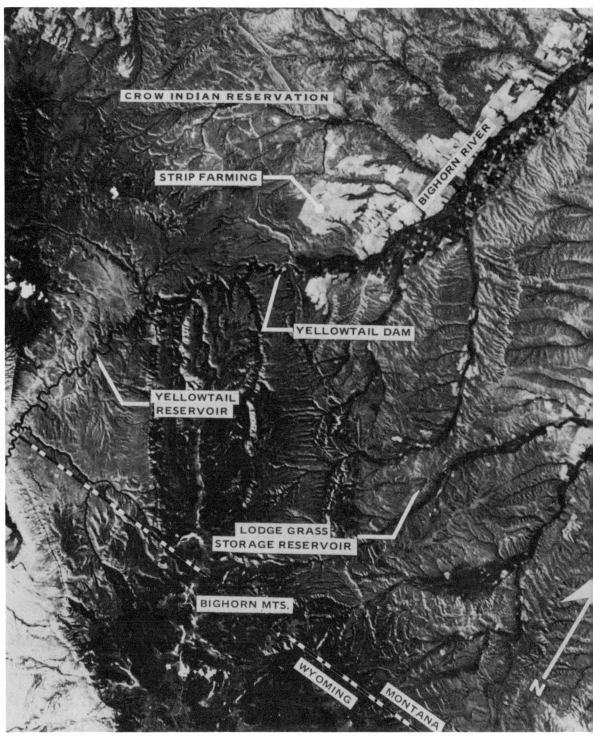

CROW INDIAN RESERVATION

STRIP FARMING

BIGHORN RIVER

YELLOWTAIL DAM

YELLOWTAIL RESERVOIR

LODGE GRASS STORAGE RESERVOIR

BIGHORN MTS.

WYOMING

MONTANA

N

The Big Horn area of Montana-Wyoming, from Skylab. Farming areas, dams, and reservoirs are revealed along with the intricate detail of the mountains and the natural watercourses that drain them. (NASA)

116

urgent meeting of grain-exporting nations to coordinate measures to deal with massive wheat shortages. Crop failures due to droughts, floods and disease were widespread, yet consumption was increasing. The result: by the end of the year global grain reserves sank to a 20-year low. In November 1974, the chief of Canada's Wheat Board called the food crisis the "worst . . . in nearly 30 years." Commissioner G. N. Vogel added that "Belts will be tightened and diets adjusted as food prices rise in virtually every corner of the world." Meanwhile, in the same month, a World Food Conference was held in Rome under United Nations sponsorship to attempt to develop emergency measures to come to grips with increasing global hunger.

Since 1961 some 50 million (of a total of 350 million) acres of farmland in the United States had lain idle as a result of farm-support programs. By 1971, as the world food crisis worsened, some of this idle land began to return to service,

and by 1974 the Department of Agriculture had permitted all of it to enter agricultural production. Yet even this dramatic move has been unable to stave off the growing threat of world famine. International health authorities predicted that before the end of 1974, between 10 and 30 million persons would die either of direct starvation or of diseases made fatal by malnutrition.

It has been estimated that by the year 1979 the annual global expenditure required to *expand* food production will exceed $11 billion. Moreover, despite the green revolution resulting from the development of superior strains of wheat and rice, the world cannot go on feeding more and more people at even moderately acceptable dietary levels without basic improvements in worldwide management. The remote sensing, repetitive overview tool may become so sharpened during the coming years that man may one day wonder how he ever got along without it.

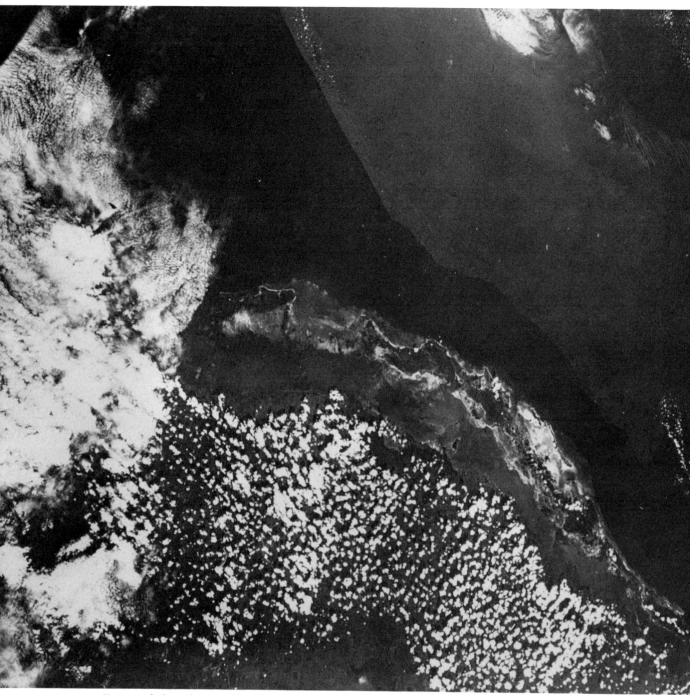

Features below the ocean's surface are revealed in this Gemini 7 photograph of the Great Bahama Bank off Cuba. Dark area at top right is the southern end of the mile-deep Tongue of the Ocean. Deep gorge of the Old Bahama Channel, running northwest to southeast, also appears darker than the shallow, 30-foot-deep waters of the banks. (NASA)

# 6

# Our Ocean World

CAPCOM:    *Skylab, Houston, we're AOS [acquisition of signal] Goldstone [tracking station],*
          *we have you for 12 minutes. We'll be doing the data/voice recorder dump here.*
SKYLAB:    *Roger . . . we got busy looking at the ground and let the time go. And the . . .*
          *ability to tell the confluence [of ocean currents] . . . we had two things working*
          *against us; number 1 was a lot of clouds and number 2 was that apparently*
          *you just can't see changes in blue, that is, from lighter blue to darker blue.*
          *Ed [Astronaut Edward G. Gibson; Gerald P. Carr was talking] could see the—*
          *the line of demarcation from where he was in the wardroom, looking out that*
          *window, but I couldn't see it . . . I did, however, see chlorophyll blooming,*
          *plankton blooming along the coast of Japan. So it's apparent that you can see*
          *the green to blue contrast but not the light blue to dark blue.*
CAPCOM:    *OK, that's a good description. Thank you, Jer.*

—DIALOGUE BETWEEN SKYLAB 4, THE FINAL
MANNED MISSION, AND THE CAPSULE COMMUNI-
CATOR (CAPCOM) AT MISSION CONTROL IN
HOUSTON, TEXAS, JANUARY 10, 1974. (THE OCEAN
CURRENTS BEING INVESTIGATED WERE THE
OYASHIO AND THE KURUSHIO.)

Planet Earth is a water planet. The oceans cover some 71 percent of its surface—140 million square miles—and it is the oceans, more than anything else, that distinguish the face of the Earth from the barren, lifeless wastes of the Moon and Mars. For ages, men have skimmed across the surface of the oceans but they have barely begun to probe their immense volume—some 330 million cubic miles of water, with an average depth of 12,450 feet. The geography of the ocean floor—with its continental shelves and slopes; submarine canyons, ridges, and basins; its abyssal trenches and globe-girdling mountain chains—is less well known than the surface of the Moon.

Orbital observations of the vast expanse of the oceans perhaps offer even more benefits than the remote sensing of the land. Aircraft can survey selected parts of the ocean surface and occasionally they are employed to interrogate radio transmitters on instrumented buoys and platforms, but costs preclude their widespread, repetitive use. Only the unmanned satellite, plus an occasional manned vehicle such as Skylab, appear to offer a cost-effective way of sensing all the world's oceans in a synoptic manner. Moreover, oceanographic satellites, like aircraft, can be used in conjunction with instrumented buoys that temporarily store information gathered by sensors on the surface as well as by subsurface sensors attached to them.

If an oceanographic satellite is placed in a synchronous, equatorial orbit, so that it remains stationary above a particular point on the globe, it will be able to monitor conditions of an entire ocean simultaneously or almost simultaneously, albeit from a very high altitude (in excess of 23,000 miles). For closer views, the same ocean can be surveyed at less-than-a-day intervals by a satellite in a fairly low (a few hundred miles) orbit.

While data from beneath the ocean surface may be relayed to satellites, the sensors aboard the satellites themselves paint essentially a two-dimensional picture, monitoring only the surface and near-surface ocean layers. This is not as great a restriction on the usefulness of oceanographic satellites as might first appear, since much of what is most significant to man occurs on the surface, where atmosphere and ocean meet. This is the area of principal energy flow that gives rise to storms, waves, and currents. It is on the surface that ships travel—and icebergs float. It is in the outer layer of the ocean that tides occur, that photosynthesis takes place, and that most of the biological resources of the sea are found. It is the

shallow parts of the ocean, covering the continental shelves, that are beginning to be exploited for oil and other minerals, and it is the shallow, heavily used, inshore waters that are polluted the most by the runoff of the land.

Oceanographic satellites can make measurements by optical, radar, infrared, and other sensing techniques of the ocean-air interface, general circulation and current changes, present and predicted surface roughness and temperature, the movement of oil spills and other pollutants, the distribution of sea ice and icebergs, and biological phenomena. As data are accumulated, they are fed into central processing centers where they are analyzed, checked against comparative "ground truth," and made available to those who need the information. In studying the oceans from space, it is essential to keep in mind that one is dealing with a dynamic (rather than static or nearly static) phenomenon: conditions at sea are constantly changing, meaning that observations and monitoring from above must be as close to continuous as possible.

Of all oceanographic phenomena, ice so far has proven to be the most readily measurable—a happy coincidence for oceanographers, since the usual "habitat" of ice is in the polar regions, where climate and distance combine to make observation by conventional methods difficult or impossible. The occurrence of sea ice and icebergs is of commercial, military, and scientific importance. Basically, users of information relating to ice need to know its present, immediate future, and probable long-term (i.e., one month) state. Among the many observables are sea ice boundaries; percentage of ice concentration per square unit of measurement; topography; thickness; and tonal variations. Others include type and age of ice; nature of meltwater puddles, cracks and other openings; drift direction and rate; time and rate of breakup of coastal ice in the spring season; and time and rate of closure as autumn and winter progress.

American and Soviet satellites have fully demonstrated the feasibility of ice surveillance in both hemispheres, revealing that the tracking of loose sea ice and icebergs has the double benefit of locating and following the ice masses themselves and of determining the velocity and direction of the oceans that carry them. (Marine biologists also take ice into consideration as they study biological productivity in far northern and southern waters, while meteorologists use it to help them make weather predictions and climatic models.) Although the satellites flown through the early 1970s were not designed primarily to study the oceans, they did point the way toward future oceanographic satellite systems and provided test beds for both active and passive sensing devices.

The Nimbus series, Skylab, and ERTS 1 craft all yielded information on polar air/water temperature and pack ice boundaries, and have discriminated between older and newer ice. Daily and weekly changes in ice concentration have been recorded, leading to practical use by the British Arctic Survey and other agencies in assisting survey ships to move safely through pack ice with minimal loss in time. Offshore leads in the Weddell Sea adjacent to the east coast of the Antarctic Peninsula were observed by Nimbus 1, 2, 3, and 5, while Essa 3 followed two huge icebergs during the 1967–1968 austral summer in the same body of water.

As a result of careful study of Nimbus 5

photography, scientists have discovered that present atlases inaccurately delineate the outer boundaries of the polar caps. Per Gloersen of NASA–Goddard Space Flight Center points out that

> The pack lines at both poles are not smooth around the ice edge . . . but consist of many indentations. For example, if you were to use a standard atlas of either pole to sail a ship into the area, you probably would be surprised to find you could also sail into a cove or channel extending into the ice pack itself. A Nimbus 5 picture would have shown where such a cove is located.

Since the satellite's radiometers can "see" through the clouds, pictures are available on a day-to-day basis regardless of weather conditions. Once the imagery is obtained it is sent to the Navy's Fleet Weather Facility (FWF) at Suitland, Maryland, for transmittal to ships operating in the polar regions. The head of FWF's Sea Ice Department, Lt. Commander William Dehn, calls such imagery "indispensable" to Arctic and Antarctic shipping operations, adding that day-night, all-weather coverage by Nimbus 5 ". . . has not only extended the navigation season, but has shown us more ice than we ever knew existed." He noted that during the late winter of 1973 the Coast Guard icebreaker *Glacier* found itself under sever cloud cover while operating in the Weddell Sea off the Antarctic coast. The availability of Nimbus 5 imagery provided the navigator with accurate information on ice boundaries of pack ice, light concentrations of sea ice, and the location of icebergs, thereby permitting him to move out into ice-free waters without danger of collision. Synoptic photographs of the

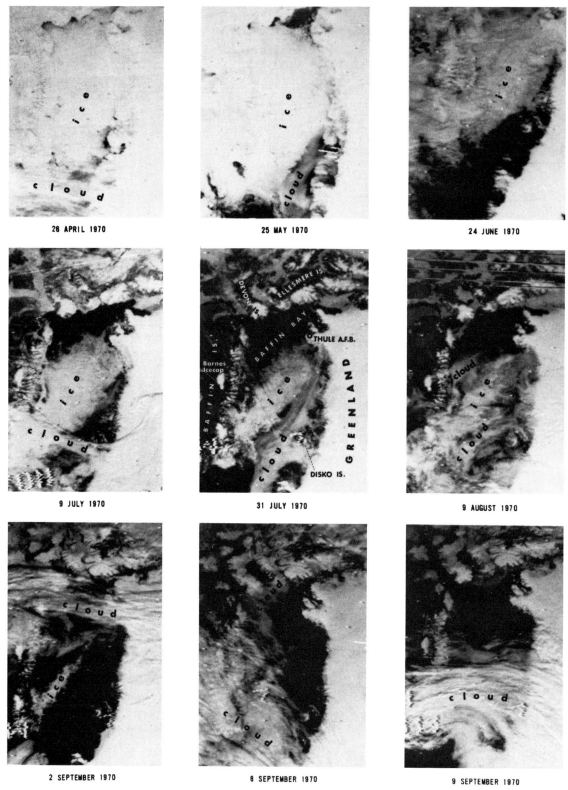

Ice reconnaissance: Break-up and gradual dissipation of sea ice from April to September 1970 as monitored by Nimbus 4. (NASA)

Antarctic coast—and of the Arctic seas also—clearly show that major changes take place over fairly short time periods.

In addition to monitoring sea ice and icebergs, oceanographic satellites can regularly supply much other basic data required by oceanographers, marine biologists, commercial and sports fishermen, shippers, and others engaged in ocean-related activities. Specifically, these craft can help reveal sea state, sea level, tides, and slope; surface and subsurface temperatures; circulation patterns—currents and surface anomalies (eddies, upwelling); water composition, and the distressing buildup of ocean pollution, much due to oil.

For example, information about the roughness or "state" of the sea traditionally is derived almost exclusively from coastal regions and along major shipping lines by ship and buoy-mounted wave recorders and occasionally from airplanes with radar and laser altimeter wave-profile recorders. Several techniques are being developed to enable unmanned satellites to provide global sea state coverage. One involves the use of a radar scatterometer (the radar return diverges from the vertical as roughness increases due to scattering by the waves). Another method under investigation involves multispectral microwave radiometry to attempt to correlate sea state with microwave emissions. Experimental work also has been undertaken, notably in the Tiros, Essa and NOAA satellite series, to relate Sun glint effects to sea state. Meanwhile, manned satellites also have provided useful information about the condition of the sea. During the Apollo 7 mission (October 15, 1968), astronauts observed vortices, slicks, swells, and characteristics that revealed current direction, internal waves, and other phenomena in the area around the Socotra and Brothers islands off East

Passive microwave radiometers reveal smooth sea conditions (left) and rough sea conditions (right). (NASA)

123

Africa. Oceanographers working with the resulting photographs stated that the wealth of information contained in them could not have been secured by conventional means.

One of the principal oceanographic results of the Skylab program had to do with the discovery of swirling eddies of cool water in the warm Gulf of Mexico currents flowing past the Yucatan. These eddies, or whirlpools, are formed by chill deep sea masses rising up into the 80°F surface waters. Robert H. Stevenson of the Office of Naval Research at the Scripps Institute of Oceanography termed the discovery "a major scientific breakthrough," one that "may change the entire thinking and understanding of thermal energy in the oceans." Adding that "This is something we could normally never have conceived," he speculated that the whirlpools would likely cause a deviation in weather forecasts, and, indeed, their very presence could possibly explain the unusual behavior of some hurricanes (that feed on heat absorbed from warm water). Although the eddies were first recorded on photographs taken by the Skylab 2 crew, Skylab 3's astronaut Gerald P. Carr—briefed to try to locate them visually—reported from orbit: "We think we saw the phenomena. We saw the cloud streaks. There's some difference in color in the water underneath, running from dark blue to light to the green."

During the last two weeks of September 1973, the second Joint North Sea Wave Project was undertaken to study the relationship between wind conditions and surface waves. NASA cooperated by sending a C-54 aircraft instrumented with an S-band radiometer, a laser profilometer, K and Ka band radiometers, a narrow

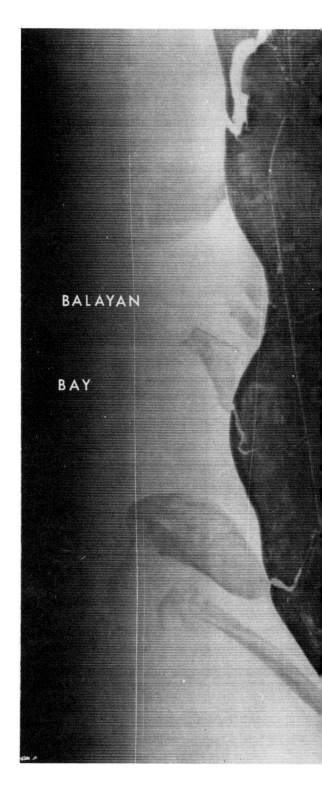

BALAYAN

BAY

124

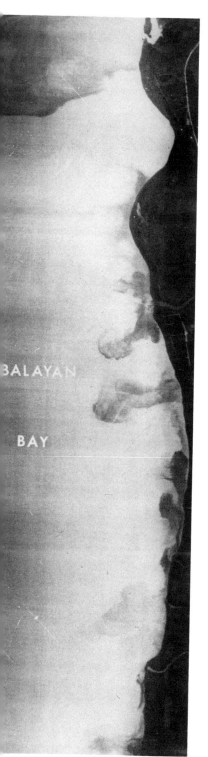
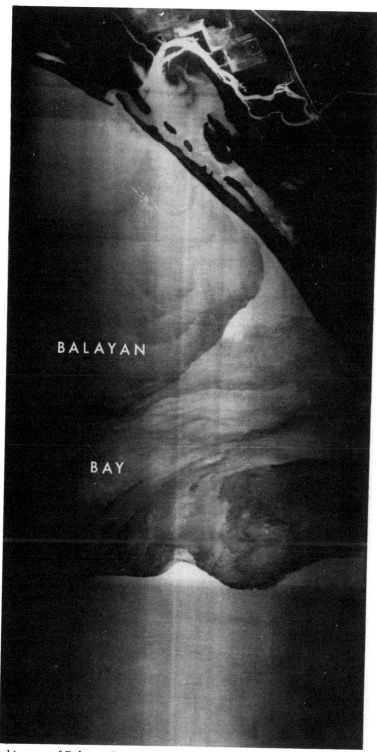

Temperature measurements: Infrared images of Balayan Bay, in the Philippines, showing plumes of ground and surface water (darker because they are cooler) discharging into the warmer, lighter-toned waters of the Bay. Photo at extreme left was made in October 1965, the other two in May 1966. (PHILIPPINE COMMISSION ON VOLCANOLOGY, USAF, U.S. DEPT. OF INTERIOR/GEOLOGICAL SURVEY)

pulse radar and a variety of cameras. During its ferry flight on September 13 from the U.S. to Europe, Atlantic sea state data were obtained in support of Skylab's radar altimeter experiment using narrow pulse radar. This was accomplished as the airplane flew under Skylab as it passed over Newfoundland and the vicinity of the Azores. During a 12-17 May 1974 symposium, the results of another joint satellite sea-study program were reviewed, the US/USSR Bering Sea Experiment on Microwave Sensing of Atmospheric and Oceanographic Characteristics. NASA reported that the study "offered convincing proof that scientists can use microwave measurements from satellites to assess sea ice distribution, motion, and stress, and that multi-spectral observations can permit scientists to determine the content of liquid water and water vapor in the atmosphere above the sea surface." The same technique was also found useful in observing how the ocean responds to surface winds.

NASA's first ocean dynamics satellite, Seasat, is instrumented to continue studies of the ocean on a global scale, including wave height and shape, the location and motion of currents, worldwide circulation patterns, winds, and the topography of the oceans. In addition, Seasat was developed with the user in mind, providing, for example, shippers with continuously updated information on sea state and ocean wind fields as well as coastal disaster warnings for shoreline populations. The satellite also is capable of observing foreign material dumped into the oceans and even providing design and siting data useful for offshore and deep water port facilities.

Temperature is another key ocean variable that may be monitored either directly by satellite-borne sensors or by combined satellite and surface techniques. The temperature of the surface and the outer layers of the ocean waters is of great importance to an understanding not only of the ocean itself but of the lower atmosphere that interacts with it. The establishment of synoptic or quasi-synoptic horizontal temperature profiles is a major goal of both oceanographers and meteorologists. Knowledge of ocean temperatures permits studies to be made of or knowledge to accrue to: (1) the Earth's heat budget (surface energy exchange), (2) weather and climate prediction, (3) circulation patterns, (4) the biological environment, (5) biological productivity (including fish distribution), (6) routing of shipping by determination of current boundaries, and (7) basic oceanographic research; e.g., large-scale dynamic thermal features of the ocean characterized by discontinuities in surface temperatures which must be synoptically and repetitively monitored. By the process of evaporation, enormous quantities of water vapor are transferred into the atmosphere, forming a major energy source in driving the global weather system. Later, water vapor energy is released by rain or snow. Oceanographers and meteorologists alike require data not only on the rates of water vapor injection into and removal from the air, but the principal areas where these actions occur.

Infrared radiometry systems flown at polar orbit altitudes of some 250 miles would be capable of making day and night measurements of sea temperatures over cloud-free areas with an accuracy of approximately 1°C and a ground resolution of perhaps five miles. The analysis of the data must take into account the fact that the infrared readings can be de-

126

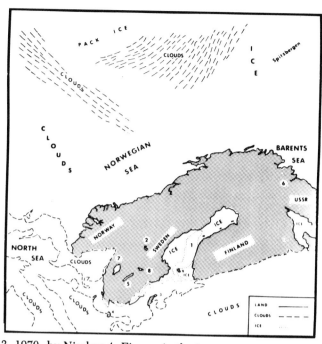

Pack ice off the Scandinavia Peninsula, viewed April 13, 1970, by Nimbus 4. Figures in the interpretation at right refer to: (1) Gulf of Bothnia; (2) Lake Siljan; (3) Gotland; (4) Bornholm; (5) Lake Vattern; (6) Murmansk; (7) Lake Vanern; (8) Lake Hjalmaren; (9) Ahvenanmaa; (10) White Sea. (NASA)

graded by water vapor radiation in the atmospheric "column" between the point on the ocean being measured and the satellite sensor. In order to prepare oceanwide surface-temperature contour maps, it is necessary to utilize surface-generated information to extrapolate the continuation of contours under cloud-covered areas. Cloud cover limitations should eventually be overcome by further development of multiple frequency microwave radiometry techniques.

Combined satellite and surface techniques have proven highly successful in the acquisition of ocean temperatures in remote parts of the world. During the 1973 austral summer, Captain Jacques Cousteau took his 141-foot-long research vessel *Calypso* to the waters off the Antarctic Peninsula to make direct read-

ings of sea temperatures and chlorophyll content. Five times a week, Cousteau's measurements were relayed via Applications Technology Satellite 3 to NASA's Ames Research Center in California where John Arveson and California State University professor Ellen Weaver later correlated them with ocean color and temperature observations from the ERTS 1 satellite and various NASA and NOAA satellites. While this was going on, data from the Nimbus 5 satellite (its microwave instruments easily penetrated cloud cover) were sent from Ames to *Calypso*, enabling her navigator to guide the ship—which had been damaged by floating ice during a storm—safely through the Antarctic ice fields to southern South America. Later, Cousteau said that without such data his research cruise probably would not have

127

Border of the Gulf Stream, as it enters the Atlantic Ocean, is shown in this Gemini 6 photograph by the line of cumulus clouds extending north from the east coast of Florida. (NASA)

been successfully completed; on one occasion, he was warned to avoid a twenty-mile-long string of ice floating in Drake's Passage.

Theoretical studies, coupled with actual experimentation with various spacecraft, show that the movements of the ocean—currents, upwellings, sinking, and the effluence of estuaries—also are susceptible to observation and monitoring from orbit. Current boundaries may be revealed by changes in water color and temperature, while the movements of the currents themselves may be observed by visible, infrared, or microwave tracking of objects (e.g. icebergs) drifting in them. Space observation of certain cloud structures, characteristic of ocean current boundaries, also can prove useful.

In February 1970, the Nimbus 3 and Itos 1 satellites both produced pictures that clearly showed the sharp thermal boundary along the north edge of the Gulf Stream; changes in boundary shape and position were noted as time went on, making it possible to prepare maps covering a two-month period. Nimbus 5 extended these observations into 1974.

Ocean upwellings also have been studied and changes in thermal patterns correlated both with phytoplankton and fish activities. In early July 1966, upwellings off the Somali Coast were recorded by the Nimbus 2 satellite; and, according to George Rabchevsky, "The development of the clockwise circulation pattern was correlated with the bloom of phytoplankton." In April 1970, the Nimbus 4 image

Changes in shape and position of the Gulf Stream's boundary off the East Coast of the United States within the space of a single month (February 1970) are evident in this series of high-resolution, infrared Nimbus and Itos photographs. Black tones identify warmest areas, white tones the coldest. At left, at start of month, boundary of Stream is close to the Coast; subsequently, it moves away, as indicated by the light-toned band of colder water between Stream and shore (NASA/GODDARD)

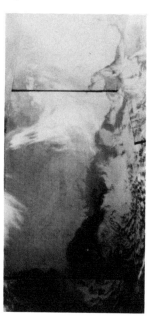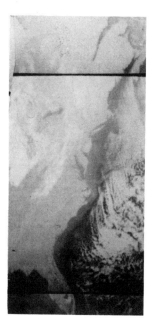

dissector camera system located another upwelling zone off southeast Saudi Arabia where the Arabian Sea and the Gulf of Oman meet. An analysis of ATS 3 satellite pictures suggested upwelling within the sun-glint patterns observed westward of the Galapagos Islands in the Pacific.

Such studies as these are potentially of great value to the fishing industry. Since fish feed on phytoplankton, and since density of phytoplankton varies on a daily, seasonal, and geographic basis, knowledge of its concentration and distribution is one of the most important factors for fishermen in determining where to head their fleets. A University of Rhode Island study of the tuna fishing industry showed that if a satellite system could reduce the time spent searching for fish by 50 percent, the per-day catch would increase by 25 percent, resulting not only in an annual savings of $15 million but a decrease of $7.5 million in investment. In other words, with the satellite observations to point the way, fewer boats would be able to do the same job more efficiently.

Paul M. Maughan of the Bureau of Commercial Fisheries has reported examples of how remote sensors aboard aircraft observe and record environmental phenomena related to fish abundance and distribution, and how they can directly detect large, near-surface schools. In the case of the latter, he notes that so far observations have only been made successfully from relatively low-flying aircraft. He also found that fish oils could be detected by multiband photometric systems. The most promising approach, however, is the indirect location of oceanographic conditions favorable to the presence of large quantities of fish. Using a Gemini 10

photograph of a portion of the Taiwan coast to illustrate the principle, he writes that:

The lighter color of the sea is the result of a diffuse sun's reflection from an evenly roughened sea surface. Winds from the northeast increased the roughness of the northerly flowing waters. The reflective, specular pattern from the sea surface thus depicts the water motion around the southern end of Taiwan.

The major current is parted by the island, much as a ship parts the water. As the "bow wave" spreads from the island, upwelling must take place near the shore. The water of darker color shown in the photograph is partly upwelled water. The presence of this water can give rise to significant fisheries. This effect is manifest in the location of the major sardine fishery of the Formosan Strait just off the western shore of Taiwan.

During the initial stages of using spacecraft for fishery research, Maughan cautions that they can ". . . provide only complementary information to that collected from more conventional platforms." Thus, we cannot expect that spacecraft will eliminate the need for ships and buoys. However, he foresees:

a global fishery forecasting network in which data collected by remote sensors will play an integral part. Earth-orbiting satellites with "phenomenon" sensors would identify the areas of the world ocean which are apparently productive. This information would be [sent] to a global fish-forecasting center

Nearly vertical Apollo 9 view of two of the Cape Verde Islands. Waves in the sun glint beyond Sao Tiago the upper island, are believed to have originated far below the sea's surface when the prevailing current met the island. Crests of such internal waves sometimes are several miles apart. Whitish area in lee of Sao Tiago indicates a calm region of upwelling, often a good fishing location. (NASA)

and analyzed, in real-time, to adjust for spatial and temporal disparities resulting from such things as variations in feeding patterns among different species of fish. Immediately, instructions would be relayed to the satellite to optically "zoom in" on an area of interest.

Coverage would be continuous, so that fish movements and migrations could be watched. Such a system would also be useful in helping international resource management and regulation of fish stocks; in fact, Maughan predicts that eventually this ". . . may prove to be as valuable as the input to the fisheries." That such a system some day may be feasible is demonstrated by the results of ERTS 1 flights over the Mississippi Sound and the northwest coast of Africa. By noting variations in water color, estimates of at least a semiquantitative nature of biological productivity were obtained on a repetitive basis.

Satellite observations of sea ice, sea state, currents, and weather have obvious implications for commercial shipping as well as for fishing vessels. From a commercial point of view, about half of the shipping bill represents costs incurred while a freighter or tanker is at sea. If, through the use of data from satellites, skippers can reduce travel time by avoiding adverse currents and areas of high sea state, it is estimated that savings for the U.S. shipping industry alone would be on the order of one-half billion dollars annually. (In a one-year test with conventional, nonsatellite routing aids, the U.S. Military Sea Transportation Service saved some $4.5 million in direct costs and $12

million in indirect costs with 1,600 co-operating ships.)

In the long run, perhaps the most important, and at the same time most melancholy, application of oceanographic satellite technology is the detection and monitoring of pollution. Some oceanographers believe that pollution, coupled with overfishing, has caused a reduction of animal and plant life in the world's oceans and seas of about a third during the past two decades. Some even warn darkly of lifeless oceans by the year 2000 if man persists in using them as garbage dumps and sewers.

As he degrades the oceans, man tampers with the very source of most of the oxygen that he breathes. Jacques Cousteau emphasized this danger to a United Nations group investigating the state of the oceans by reminding its members that "Phytoplankton, the primitive plant life that generates most of the Earth's oxygen, is surface matter. It absorbs dirt and acts as a sort of pollution filter. Thus, all you need to knock out is the surface phytoplankton and the entire marine life cycle is fatally disrupted." Measurements show the dissolved oxygen content of sea water to be diminishing; at one station, in the Baltic Sea, water samples contained 2.5 cubic centimeters per liter in 1900, 2.0 in 1940, and 0.1 in 1970.

It is not known how much oil is dumped each year into the oceans and seas, but educated estimates range from 4.5 to 10 million tons. The oil may remain in the ocean environment for several weeks, or for as long as fifty years. Where does most of the oil come from? One expert offers the following breakdown:

| | |
|---|---|
| automobile exhaust emissions (precipitated onto oceans with rain) | 1.8 million tons/year |
| spills from tankers | 1.0 million tons/year |
| dumped into oceans by rivers | 3 to 7 million tons/year |

Sooner or later, the major industrial and motor-vehicle-using countries in the world will have to combine forces to eliminate the oil (and other) pollution that some day could destroy us all. Meanwhile it appears that only satellites offer the capability of effectively monitoring the appearance, movement, and eventual dissipation of pollutants in all corners of the global oceans.

Satellites would typically make a regular examination of the buildup of pollution in the Mediterranean. As it turns out, relatively clean waters flowing in to the sea through Gibraltar are soon polluted by inflow not only from the Adriatic and Aegean but from such rivers as the Rhone and the Nile. The Red Sea, connected to the Mediterranean only by the Suez Canal, is quickly dying. Cousteau predicts that soon "there will be nothing alive" in its depths, a fate that larger bodies may one day experience unless aggressive corrective action is taken by man.

Although the only feasible way to monitor effectively water pollution on a wide scale is by orbital surveys, interim efforts have been made to establish water quality measuring stations on the surface. In New York, for example, twenty-two such stations are tied together by telephone lines to record pollution levels in different localities and then to feed the data into a central computer in the state capital at Albany. The continuous flow of information permits the computer to determine where pollution levels are rising significantly, helping environmental experts to identify the sources. This program illustrates an important step toward a nation or even worldwide pollution monitoring system wherein surface monitors feed their measurements directly to data relay-satellites as they pass overhead.

Instruments developed for NASA's experimental Nimbus G program are designed to measure the characteristics of polluted water as it is driven across the oceans by surface winds as well as the cooling of the oceans by arctic air. Thus, the satellite will be able to observe simultaneously the two reservoirs of pollution on our planet—atmosphere and ocean—and the processes they undergo as they distribute the pollutants they are increasingly forced to carry.

Water pollution is present in many shapes and forms, in one way or another due to man's increasing need to dump the by-products of his civilization. Sometimes enlightened and active groups can prevent ocean dumping, but more often it goes on either because no one observes it or because nothing can be done about stopping it. A typical detected case involved the Dutch ship *Stella Maris*, which was chartered by a large chemical firm in the summer of 1971 to dump some

133

Oil spills are readily detectable by overview imagery. This photograph, taken originally in color infrared, shows oil and sediment resulting from the March 1969 Santa Barbara oil spill off California. (U.S. DEPT. OF INTERIOR/GEOLOGICAL SURVEY)

600 tons of chlorified aliphatics hundreds of miles out at sea. It was, however, prevented from dumping by strong protests from the nations that might have been affected; the British called the *Stella Maris* the "ship of shame," and a small Irish destroyer followed her closely. Finally, the ship went back to its home port of Rotterdam with the dangerous chemicals aboard, transferred them to storage, and awaited other tasks—principally dumping more of the millions of tons of chemical (including radioactive) wastes flowing from the great German, French, Dutch, Belgian, and Swiss industries that rely on northwest European ports.

At such time as world maritime nations frame agreements not to dump dangerous chemicals into the oceans it will become necessary to establish a method of global monitoring, of following ships like the *Stella Maris* (which changed its name, incidentally, to *Constance*) as they move out to sea, and of detecting by remote sen-

Fire on a drilling platform, off the Louisiana coast in March 1970, and the resulting oil spill, seen during cleanup operations. (U.S. DEPT. OF INTERIOR/GEOLOGICAL SURVEY)

sors the chemical wastes they (presumably illegally) throw into the oceans.

In this connection, an interesting new technique for monitoring oil pollution from the surface has been devised by the General Electric Company's Research and Development Center. Before an oil tanker sets off to sea, GE proposes, its petroleum cargo would be tagged with a small batch of magnetic ferrites in the form of dust. Each tanker's dust would be slightly different from that carried by others, so

should a spill occur the magnetic tags in the oil would reveal the guilty vessel and its owners would be assessed the cleanup costs. Extrapolations of this or similar techniques may in the future be tied to oceanographic satellites, resulting not only in the nearly instantaneous observation of an oil spill but the rapid identification of the cause. The only apparent alternative would be some sort of United Nations navy that could shadow dump ships, perhaps firing across their bows if

135

Cape Hatteras, North Carolina, from Apollo 9; south of it, not jutting out as far into the ocean, is Cape Lookout. Sediment passes from Pamlico Sound, between the two capes, and from smaller Albermarle Sound, to the north, into the Atlantic, where the dirty waters meet and define the boundary of the Gulf Stream. (NASA)

This Gemini 5 photograph of Laguna de Terminos, Campeche, Yucatan, Mexico, reveals a number of coastal processes at work, including long shore and offshore currents. A river drops marl, silt, and sand sediments into the clear waters of the Gulf of Mexico, revealing the currents. (NASA)

they started to unload their nefarious cargos.

Water pollution is greatest, of course, where there are the most people—and that means along the coasts. In the continental United States, about one-third the population inhabits the 15 percent of the land represented about the coastal regions. Some two-fifths of all U.S. manufacturing is located along the coasts, with chemical, petroleum, paper, and primary metals activities having the greatest impact upon the waters of the estuarine zone. This zone, generally defined as the narrow area between the landward limit of tidal influence and three miles to seaward, actually varies geographically depending upon the extent of freshwater runoff. In parts of California, the zone is only a few yards wide, while off the Mississippi delta it extends for fifty miles —and even further off the mouth of the Amazon. What happens in and to the waters of the estuarine zone is of especially great importance to mankind because, as the U.S. Department of Interior's "National Estuarine Pollution Study" put it:

> While no life form can be singled out as irreplaceable, the kinds of life which need the estuarine zone to survive represent essential links in the energy conversion chain upon which man depends for survival. Many of the human uses of the estuarine zone depend directly or indirectly on the existence of the estuarine zone as a healthy habitat.

As with other aspects of Earth resources management and protection, the overview is unquestionably the most feasible tool to apply to estuarine problems. Synoptic observations of many national, and world, coastal areas are made by specially equipped aircraft and spacecraft. Sensors assist in measuring water properties, including the buildup of pollutants, by taking advantage of such observables as surface temperature and compositional boundaries of effluents (which may indicate stagnation in estuary inputs from polluted rivers, and thermal pollution). They also detect colors (which reveal chemical pollution, polluted fresh water inflow, and pollution possibly caused by nuisance algal and weed growth) as well as unusual chlorophyll content (indicative of some chemical and biological pollution possibly caused by insecticides and agricultural chemicals). The U.S. Public Health Service is almost certain that the pesticide Endrin was the agent that destroyed 10 million fish in the Mississippi River–Gulf of Mexico brackish water zone in 1964, a single example of havoc caused by man on estuarine life.

The seriousness of the coastal and estuarine zone situation is underscored by the Department of Interior's estimate that three-quarters of the entire U.S. tidal shoreline, which is 53,677 miles long, suffers from moderate to severe modification due to the hand of man. Long Island Sound was twice as polluted in 1974 as it was a decade ago, gravely damaging its aesthetic, recreational, and shellfish harvesting qualities. One-third of San Francisco Bay has been covered with landfill, the water is polluted, and housing and other construction cover virtually every inch of waterfront. New York harbor contributes as a prime barge dumping area of chemicals, and is thoroughly infested with sewer sludge. ERTS 1 experimenter C. T. Wezernak of the

An example of how the overview helps in monitoring waste disposal in the ocean. Here, sewer sludge and acid wastes build up in New York Bight. (NASA)

University of Michigan confirmed, as a result of multicolor satellite imagery, that sewage and industrial acid wastes were pooling just outside the entrance to the harbor. By observing the geometric configuration of the waste dumped into the harbor, the products responsible for altering water color were determined. Moreover, it was possible to track their dispersal by tidal currents.

The satellite overview also shows how estuaries and wetlands along the coasts are being filled or otherwise modified at an alarming rate. The Department of Interior estimates that some 650,000 acres were destroyed, ecologically speaking, by fill and dredging operations in the 1950s and 1960s. Moreover, over a quarter of

America's 1.4 million acres of shellfish area is already polluted and unless corrective action is taken, all will be destroyed. According to Edward Yost of Long Island University, an ERTS 1 photo taken late in July 1972 of southern New England showed the presence of larger-than-normal amounts of sediment that appear to have been related to an onslaught of poisonous "red tide" plankton. This observation led to the cessation of all shellfishing in the region at a cost of about $1 million per week. Not only is this a cultural tragedy, but the protein food value of shellfish is increasingly needed as world population soars.

Many of New England's rivers are severely polluted, and the 20,000-square-

139

The New England coast, from Skylab 2, including the eastern half of Long Island, below, and the distinctive fish-hook of Cape Cod, just above the islands of Martha's Vineyard and, farther to sea, Nantucket. The New York border with Connecticut and Massachusetts falls close to the lefthand margin of the photo, while the northern border of Massachusetts runs along the photograph's top. (NASA)

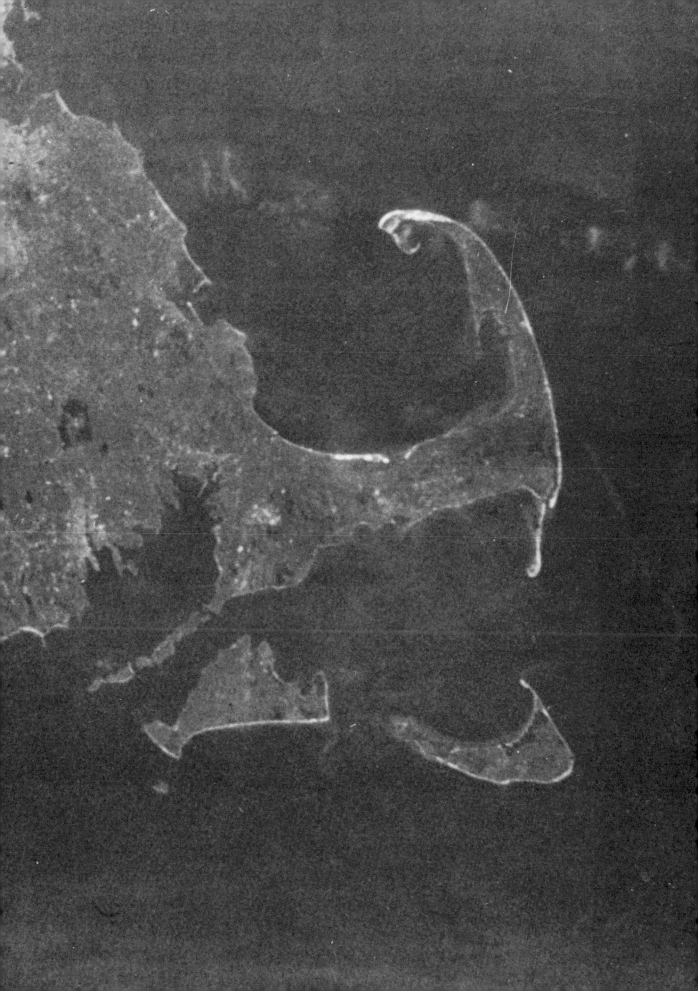

mile Chesapeake Bay is beginning to feel the effects of contamination as the Susquehanna pours in its offal. In Savannah, one chemical company alone throws nearly 700,000 pounds of sulfuric acid daily into the Savannah River. Most of Texas's 3,350-mile tidal shoreline is disturbed by emissions of oil refineries and chemical plants, and the 50-mile-long Houston Ship Canal is in extremely bad condition.

Conditions are not quite so bad along the coasts of northern California, Oregon, and Washington where riverine and estuarine systems are less complex than in the East and fewer harbors attract fewer ships and less industry. But Washington's Bellingham Bay already has been heavily damaged by paper mills and developers have endless plans for further use of the area that inevitably will lead to environmental degradation.

Thus, the overview of the estuarine zone is a grim one, but perhaps by virtue of it, measures can be implemented to curb its pollution.

In addition to detecting and monitoring sources of pollution, the satellite overview of the estuarine zone is extremely useful in a number of other respects. Satellites can be used to monitor the heavy shipping traffic that converges near ports, to warn of the approach of hurricanes and tidal waves, and to map the ever-changing shorelines whose dynamic nature has long thwarted efforts to maintain charts that are *current* as well as accurate.

The U.S. Army Corps of Engineers has found orbital imagery—particularly the repetitive type emerging from the ERTS 1 program—to be of inestimable value in keeping up with changing coastlines. Outside the United States, a single ERTS 1

Rapid changes in coastline can be monitored by satellite. This Gemini 4 photograph of the Florida Bay area and nearby Keys shows configuration below the surface. Vegetated and nonvegetated shoals are evident, along with a live reef parallel to the Keys, as depicted in sketch below. (NASA)

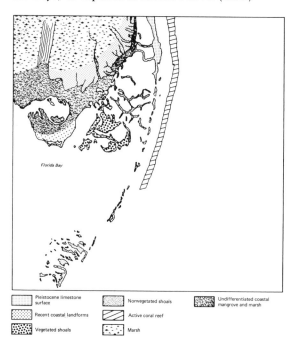

Pleistocene limestone surface

Recent coastal landforms

Vegetated shoals

Nonvegetated shoals

Active coral reef

Marsh

Undifferentiated coastal mangrove and marsh

Florida Bay

image of part of the Australian coastline showed charts to be gravely outdated. Experiments with sensors aboard ERTS 1, Skylab, and other satellites also have demonstrated the feasibility of observing the changing nature of offshore islands along the Atlantic and Gulf coasts of the United States. Storms and the hand of man affect the characteristics and stability of these islands, which are significant in that, on the one hand, they create hazards to navigation, while on the other, they provide harbors for small boats, land for real estate development, protective barriers to the shoreline, and habitats for coastal birds and marine life.

Still, for the estuarine zone and perhaps for the oceans as a whole, the satellite overview is likely to have its greatest value as a means for controlling pollution. The suggestion that man might one day be forced to protect the oceans from his own mistreatment would have sounded incredible even a few decades ago. For most of history, the ocean stirred up images of fear, respect, pleasure, or nostalgia—not concern for its survival. The feeling for and sound of the sea were forever being invoked by the nineteenth-century poet, Epes Sargent:

A life on the Ocean wave,
   A home on the rolling deep;
Where the scattered waters rave,
   and the winds their revels keep!
Like an eagle caged I pine
   On this dull, unchanging shore!
Oh, give me the flashing brine,
   The spray and the tempest's roar!

Today's poet must strike a different chord. For now, the oceans have assumed a new importance for man. Living, healthy and vibrant, they will help to assure his survival and progress toward a future of prosperity and happiness. Polluted and lifeless, they will spell his doom, and an experiment of nature that began billions of years ago when man's remote ancestors arose in the seas will end in failure.

Cirrus clouds over the Atlantic, northwest of Dakar, on Africa's westernmost tip. Transverse bands may indicate a spiraling circulation within the subtropical jetstream that sweeps north over the Sahara to the Mediterranean. Photo was taken by remote control during the unmanned Apollo 6 flight. (NASA)

# 7

# Looking Down at the Atmosphere

*Looks like there is a fairly strong mass of polar air moving from the southwest up towards Tierra del Fuego. It's mixed with some cloudiness that extends from that area all the way down to the Antarctic ice shelf . . .*

*There is cloudiness along the Andean ridge and also in the Amazon basin, stretching from the eastern coast of South America on up, oh, two-thirds of the way towards Central America. It doesn't look like frontal weather there. It's probably tropical convergence weather.*

*The eastern half and the midwest of the United States is completely cloud-covered by now.*

*The Pacific regions, the west coast of the United States is cloudy—at least west of southern California. I see no strong new frontal patterns . . . There may be one that would be lying maybe across . . . northern California and into Colorado, with a little clear area ahead of it, possibly in Kansas, but then into this solid bank of clouds that stretches from Brownsville along the Gulf Coast across the panhandle of Florida, up the east coast on out to Nova Scotia . . .*

—ASTRONAUT HARRISON H. SCHMIDT, APOLLO 17, DECEMBER 8, 1972

The potential uses of the satellite overview of the Earth's continents and oceans have only begun to be exploited but in the case of atmospheric research, the situation is quite different. Here the artificial satellite already has proven indispensable to the global study and forecasting of weather. Since the launching of the first Essa satellite in February 1966, the world's weather has been monitored daily from space and hundreds of major typhoons and hurricanes have been tracked. The use of satellite photos has become a part of the daily routine, not only for weather forecasters but for commercial and military pilots who regularly check them before takeoff.

The atmosphere, like the ocean, is ever-changing; events in one geographic area often affect those in another. Ground observations, even in populated regions, can never be pieced together with the same thoroughness and rapidity as from orbit. And vast expanses of land and ocean surface exist where no meteorological stations are located and from where data are only obtained occasionally, as from a passing ship, a scientific exploration party, or an emplaced automatic weather monitor such as the Expendable Remote Operating Weather Stations used by the U.S. Air Force.

Recognizing that weather can only be studied and predicted globally, the member nations of the World Meteorological Organization (WMO) created the three-part World Weather Program: the World Weather Watch (WWW), the Global Atmospheric Research Program (GARP), and the Systems Design and Technological Development Program. The origin of the program goes back to December 20, 1961 when a U.N. resolution called upon WMO to develop measures to advance our knowledge of climate in general and weather forecasting and weather modification in particular. The overall goal set by WMO is the gathering of data and the conducting of research on the physics and the dynamic processes of the atmosphere in order to develop the knowledge and techniques necessary to make accurate weather forecasts of at least a week and ultimately two weeks in advance. Other objectives are to study, with mathematical models and computer techniques, means of modifying weather at critical times and places, and to further cooperation in meteorological research (including the dissemination of weather information).

The principal objective of the WWW element of the program is to develop an atmospheric observation and weather service covering the entire Earth and to assure that information is processed,

analyzed, and disseminated promptly. GARP, a joint project of WMO and the International Council of Scientific Unions, involves the undertaking of basic research into atmospheric physics and the planning and implementation of gathering the data necessary for such research. For example, one GARP activity is concerned with the *planetary boundary*, or the interface between the oceans and the first few thousand feet of air. Energy is transferred to and from both media within this layer, and hence an understanding of its nature is absolutely essential to comprehension of the workings of the atmosphere as a whole. A specific project in this study was code-named BOMEX, from Barbados Oceanographic and Meteorological Experiment. BOMEX took place in May, June, and July 1969 and involved seven agencies of the U.S. government, the government of Barbadoes, and more than twenty-five universities and research institutes—scientific personnel exceeded 1,500. Ten ships, twenty-four aircraft, a number of meteorological satellites, and twelve buoys were employed in the experiment, whose principal objective was to determine the quantity of moisture and heat energy released by the ocean into the atmosphere over a 90,000-square-mile tropical area.

Large-scale experiments of the BOMEX sort were preceded by years of research in high-altitude cloud photography, with aircraft, balloons and, after World War II, with V-2, Aerobee, Viking, and other sounding rockets. Of particular interest was the discovery in October 1954 of a tropical storm in the Del Rio, Texas, region that had not been observed by conventional methods. Rocket flights underscored the potential of high-altitude photography, but did not provide meteorologists with the regular, repeated, synoptic observations they needed for analyses of weather. Accordingly, with the impetus to satellite experimentation given by the International Geophysical Year of 1957–1958, several experiments involving Earth heat-budget measurements and cloud-cover mapping were carried out toward the end of the 1950 decade.

During the 1960s, the United States put into service a number of meteorological satellites, including ten Tiros satellites that had lifetimes of from 89 to 1,809 days and nine Essa craft, the last of which was orbited in late February 1969. Some of these satellites carried Automatic Picture Transmission (APT) equipment designed to send vidicon images of the cloud cover and of the exposed surface to numerous ground receiving stations. Other satellites were fitted with Advanced Vidicon Camera Systems (AVCS), whose vidicon tube images were first scanned and then converted to signals that were stored on magnetic tape until time for transmission to either of two central command and data acquisition stations in Alaska and Virginia. At the National Environmental Satellite Center in Maryland, maps made from the data obtained from these satellites were— and still are—sent to recipients all over the world (including Russia, via a direct circuit from the World Meteorological Center in Washington to another in Moscow).

Tiros and Essa satellites were followed by Itos 1, the first of the improved Tiros operational satellites, and its successors carrying the NOAA designation (for National Oceanic and Atmospheric Administration of the Department of Commerce). These newer satellites provide

more efficient daily weather coverage at lower cost than their predecessors. A much improved altitude control system assured that their sensors and antennas always pointed downward to the surface, not just part of the time as occurred in the earlier, spin-stabilized craft that traveled with a fixed orientation as they spun (their sensors and antennas sometimes pointed right at the Earth, sometimes obliquely away, and sometimes out into space). NOAA 2, launched on October 15, 1972, from the Western Test Range at Lompoc, California, was the first satellite to employ operationally a very high resolution radiometer capable of taking pictures of cloud tops and land surface from which temperatures can be determined. NOAA 2 covered the world's atmosphere and oceans twice a day as it orbited 900 miles above the Earth every 115 minutes. It was followed by NOAA 3, orbited on November 6, 1973.

All U.S. weather satellites were developed for the National Oceanic and Atmospheric Administration by NASA, which continues to advance the state of the art with its experimental Nimbus series. Carrying APT and AVCS devices, these craft also have tested such instruments as the filter wedge spectrometer that can monitor the distribution of water vapor in the atmosphere, and the temperature and humidity infrared radiometer that measures infrared radiation from the Earth (making possible the construction of maps of day and night cloud cover, and of cloud, land, and ocean-surface temperatures). This instrument also monitors the content and contamination of cirrus clouds and the relative air humidity. Another instrument, the backscatter ultraviolet spectrometer, measures the spatial distri-

Satellite photography of Earth's western hemisphere shows at a glance storm centers in otherwise sparsely monitored regions of the Atlantic Ocean, as well as a tropical disturbance and, bracketing South America, two fronts. (NOAA)

148

Composite view of the northern hemisphere's meteorological conditions revealed by weather satellites operated by the National Oceanographic and Atmospheric Administration. Only the extreme polar regions are not covered by satellite imagery. (NOAA).

149

bution of ozone in the air by observing the intensity of ultraviolet radiation backscattered from the atmosphere. And the selective chopper radiometer records the temperature of six different, six-mile-deep layers of air. Nimbus 5, orbited on December 12, 1972, carried six instruments, two of which were improvements over types carried on earlier spacecraft and four of which were new. Both the electrically scanning microwave radiometer and the microwave scanner utilized, for the first time from space, the microwave region of the spectrum for remote sensing. Microwaves are electromagnetic waves of very high frequency that can readily penetrate cirrus and stratus clouds that are opaque to the passage of infrared radiation. The use of microwaves thus gave Nimbus 5 the ability to sense effectively on cloudy days. Another new instrument, the infrared temperature profile radiometer, measured radiation from the surface and from the atmosphere in seven spectral bands, while the surface composition mapping radiometer probed for differences in thermal emissions from the Earth below. Like NOAAs 2 and 3, Nimbus 5 observed the entire planet twice every twenty-four hours, once in sunlight and once in darkness.

NASA's first Synchronous Meteorological Satellite, SMA-1, was launched on the 17th of May 1974 into a 24-hour geosynchronous orbit over the Atlantic Ocean. This pioneering weather observation spacecraft transmits electronic data from which hemisphere-wide daytime and nighttime pictures are constructed. These pictures, emerging at the rate of one each half hour, can be received by hundreds of small receiving stations located principally in the Western

Hemisphere. In addition, SMA-1 receives and transmits environmental data from nearly 10,000 manned and unmanned data collection platforms as well as some 20 high-flying balloons. Also on board is a space environment monitoring system, used to measure solar activity and its effect on the terrestrial environment.

Meteorological studies also have been made by non-weather satellites, notably Dodge (that took the first color picture of the full Earth on July 25, 1967) and the Applications Technology Satellite series that dates from the 1960s. ATS 1 and ATS 3 carried spinscan cameras that produced photographs of the entire disk of the planet each twenty minutes, enabling meteorologists to prepare time-lapse motion pictures of changing cloud patterns. Located in a stationary geosynchronous orbit 22,300 miles above Colombia at 70 degrees west longitude, ATS 3 occasionally suffers difficulties from excess heating when the Sun is north of the equator in summer. Normally, the 805-pound spacecraft spins at 100 revolutions per minute in one direction while the antenna spins at nearly that speed in the opposite—when coupled with the satellite's orbital motion, the antenna is thus kept pointed directly at the Earth below. In mid-July 1971, antenna speed dropped to 80 then 50 rpms, briefly resumed full speed, and then in early August, the drive mechanism stopped completely. National Oceanic and Atmospheric Administration weather scientists (particularly those at the National Hurricane Center in Miami) were keenly disappointed, as they had relied heavily on this multi-purpose satellite for photographs of Western Hemisphere cloud cover and storms. The problem began to clear up, however, in early September as

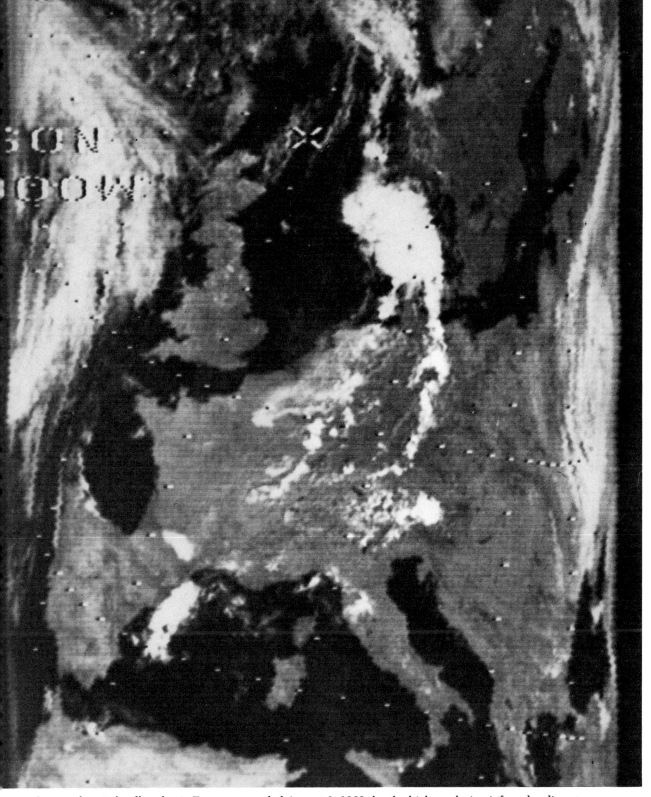

A rare, almost cloudless day in Europe, recorded August 6, 1969, by the high resolution infrared radiometer aboard Nimbus 3. (NASA).

the Sun moved south and the control mechanism cooled off.

The latest in the ATS series, No. 6, orbited on May 30, 1974, carries a very high resolution radiometer experiment that provides both visible and infrared images of day and night cloud cover over about one-fifth of the Earth. The principal purpose of the experiment is to observe cloud motion, storm life cycles, and such small-scale phenomena as thunderstorms.

In Russia, a number of Kosmos and Meteor satellites have been instrumented for meteorological experiments, and such non-weather satellites as Molniya have taken pictures of the planet, from very high altitude, that depict weather condition on a continental scale. The French Eole satellite, launched by a U.S. Scout rocket at Wallops Station in Virginia in August 1971, was designed to receive and relay wind, temperature, and pressure information being gathered in the Southern Hemisphere by up to 500 instrumented balloons flying at 39,000-foot altitudes. These balloons, launched from Argentina by Centre National d'Études Spatiales and local personnel from three different sites, could be interrogated at any time, day or night, alone or in groups of up to sixty-four. Excellent weather information has also been gained from the U.S. Mercury, Gemini, Apollo, and Skylab manned spacecraft and the Russian Vostok, Voshkod, and Soyuz series. Moreover, deep-space and lunar probes have produced fascinating facsimile pictures, including the first photograph of Earth from orbit around the Moon, taken by Lunar Orbiter 1 on August 23, 1966. With the advent of the manned Apollo flights Moonward in December 1968, stunning color film of our planet began to

be returned, clearly showing the brilliant white clouds that cover so much of the world.

Hardly any of man's activities escape the effect of weather and some are especially weather-sensitive, such as sports, recreation, agriculture, fishing, natural resource extraction, construction, air transportation, filmmaking, and shipping. Weather, of course, can be destructive as well as simply inconvenient. The strongest hurricane ever to devastate the United States—Camille, which hit the Mississippi coast in 1969—packed winds of up to 200 miles per hour; it killed more than three hundred persons and cost an estimated $1 billion in property damage. It is impossible to know exactly what the loss of life and property might have been had not advance information on the storm been available from weather satellites, but meteorologists have estimated that without this information, 50,000 persons might have perished. Later in the same year, satellite tracking of hurricane Laurie indicated that it would not strike the Gulf Coast, and hence that protective and evacuation procedures would not be necessary; this meant a savings of more than $3 million that otherwise would have been spent.

One does not have to be in the path of a killer cyclone to realize that improved weather prediction will produce enormous benefits to humanity. The U.S. National Academy of Sciences has estimated—conservatively—that better long-range forecasts would save the United States some $2.5 billion annually by minimizing flood and storm damage as well as the adverse effects on new construction, electric power generation, and fruit, vegetable, and livestock production.

The Academy did not try to judge the number of lives that already have been saved or the value of the property around the world that is safeguarded because of weather satellite information. The magnitude of the savings might best be appreciated by singling out one small example. According to a study by the University of Wisconsin, weather satellite data will save the state's farmers between $30 million and $35 million annually during the next decade—and this is on their hay crops alone.

Weather satellites also are bound to play an increasing role in the detection and monitoring of air pollution on a global scale. The seriousness of the pollution problem now varies from country to country and from place to place, depending on the level of industrialization, weather patterns, the nature of the terrain, and the effectiveness of the antipollution measures being taken. There is some evidence that motor car pollution is beginning to level off due to the gradual introduction of antipollution devices. Moreover, the reduction or elimination of coal and waste burning in some communities has led to lower solid-waste emissions. Nevertheless, the overall environmental problem worsens, though the extent that it does so is hard to evaluate due to the lack of sufficient statistical data. Such data can, to an extent, come from ground measurements, but these provide only local relief to an extremely difficult situation. Even though thousands of air quality checking stations may become available, the information they could provide for continent-wide statistics would be marginal at best. Only the continuous orbital overview can provide the data necessary to monitor conditions

as they are and to predict what they may become in the near future. (ERTS 1 successfully relayed a large number of data messages from thirty-three environmental sensing platforms established by the U.S. Geological Survey, graphically demonstrating the usefulness and flexibility of the Earth resource satellite as a relay communications system.)

Many difficult questions must be solved before satellites can assist in mounting a truly effective attack on global air pollution but continuing experiments with satellites are helping to clarify the basic requirements for such a reconnaissance system. For example, detection of pollutants in the lower atmosphere is complicated by the fact that they generally concentrate in thin layers close to the ground rather than rising and mixing at higher levels. This means sensors will have to be sensitive to low-layer contaminants, and not be confused by substances found elsewhere along the satellite-to-ground observation path. In order to monitor effectively the sources of air pollution, the intensity and movement of the effluents, and the nature of the individual constituents, a combination of satellite and ground detection systems will be necessary to provide the information on which maps can be constructed showing once or twice a day pollution flow from municipal and industrial areas. Pollution monitoring satellites will not only employ their own sensors but serve as relays of data transmitted up from the ground stations. Such satellites could eventually provide policing and prediction functions as well as detection and reconnaissance. Predictions of local and regional pollution accumulations and their intensity would be invaluable so that preventive measures

(continued on page 162)

153

Dynamics of the atmosphere: The fluidity of the atmosphere is demonstrated by the below Apollo 9 view of the Cape Verde Islands; each cloud streak in center is caused by an island interrupting the northeast trade winds, while alternative semicircular cloud lines in distance result from interweaving of oppositely curving wind currents around islands. Right, a Gemini 5 view of trade wind cumuli, arranged in polygonal cells, rows, and forked lines east of the Lesser Antilles. (NASA)

155

Left, the Pacific Ocean, with 90 percent cloud cover, viewed from Apollo 7. Concentric rings of large storm cell are the result of widening rims of successive central updrafts in the cell. Above, the chain of vortices, similar to eddy patterns, stretching downwind from the island of Guadalupe in this Gemini 6 photograph of the eastern Pacific Ocean, indicates cellular convection occurring in air near ocean surface. Guadalupe itself is surrounded by stratocumulus clouds (NASA)

157

The eye of the storm: Left, a vertical view of 1973's hurricane Ava, by the NOAA 2 satellite; right, hurricane Ellen, over the Atlantic in 1973, from Skylab 3; below, a family of killer tornadoes that hit the Midwest in April 1974 was spawned by the weather system seen in this Defense Meteorological Satellite Program photo. Tornadoes developed just ahead of the front of warm moist air being pulled northeast from the Gulf of Mexico by a low pressure area. (NOAA/NESS, NASA, USAF)

Monitoring pollution: Above, the Los Angeles Basin, from La Jolla to Santa Barbara, showing incursion of air pollution inward to Palm Springs area; offshore islands are San Nicolas (left), San Clemente (below), and Santa Catalina (above). In Europe, right, heavy air pollution obscures portions of Germany, Holland, and Denmark. (NASA)

160

Gulf of Mexico, south of Louisiana, reflects early morning sunlight as, to left, smoke plumes pour into air in Houston-Galveston area. Near center, above Shreveport, a jet aircraft contrail can be seen. Bright lines in eastern Louisiana and Mississippi are produced by ground fogs. Thunderstorms are building up over north Texas as a cold front moves southward. (NASA)

could be taken, such as the shutdown of industrial plants and the control of motor traffic. As a partial, interim solution, NASA announced in mid-March 1973 that it would equip a Boeing 747 aircraft to monitor pollution in the upper atmosphere as part of its Global Air Sampling Program. Instruments selected are capable of detecting gaseous pollutants and of measuring the number, size, and distribution of airborne particles.

On a longer-term scale, satellites will provide surveillance of the global buildup of atmospheric pollution, not only in the lower atmosphere but in the upper troposphere and stratosphere as well. Realizing that the increase of carbon dioxide, for example, would tend to cause rises in worldwide temperatures, while increasing particulate matter suspended in the air would tend to lower temperatures, meteorologists will want to determine what long-range effects man is having on his climate. And what permanent, perhaps irreversible, damage he is doing.

Less familiar than the cloud and weather patterns of the lower atmosphere is the outer atmosphere into which pour streams of charged particles from the Sun. Directly or indirectly, these affect the reliability of satellite instruments, the safety of man in space, and radio communications. The outer atmosphere-border of space environment is analogous to lower atmospheric weather, and hence can be thought of as "space weather." Any disturbance to this weather is a "space disturbance." Various spacecraft of NASA (Pioneers, Explorers, Applications Technology Satellites, and Orbiting Solar Observatories) and the Department of Defense (particularly Velas and Solrads) provide regular monitoring of solar flare activity. A Space Disturbance Laboratory was established in Colorado to predict and to monitor fluctuations in the radiation environment of the upper atmosphere and near-Earth space and to conduct basic research on the interactions of solar protons and the upper atmosphere, the interaction of the solar wind with the magnetosphere, and particles trapped in the Earth's magnetic field. Many of the data needed for monitoring space disturbances and making forecasts are gathered from instrumented satellites. Disturbance information is needed not only for vehicles like manned Apollo and Skylab spacecraft that travel beyond the sensible atmosphere but for aircraft that may fly high enough for solar flare-atmospheric interreactions to pose hazards to crew and passengers.

Meanwhile, so-called topside sounders, or satellites traveling at altitudes of about 600 miles, are employed to probe the electron density of the upper ionosphere from above. Other satellites are needed to map the ionosphere from within, orbiting at altitudes between 120 to 200 miles. Among the ionospheric sounders are the Alouette series of satellites, Explorer 20 and Isis 1. Still additional vehicles are instrumented to investigate natural and man-made radio noise and interference so that global maps can be constructed showing contours at satellite altitudes for frequencies between 100 kc/s and 100 mc/s. Lightning discharges all over the world would also be recorded. An appreciation for the economic potential from satellite aeronomy research can be gained when it is realized that the U.S. telecommunications industry alone spends more than $20 billion a year.

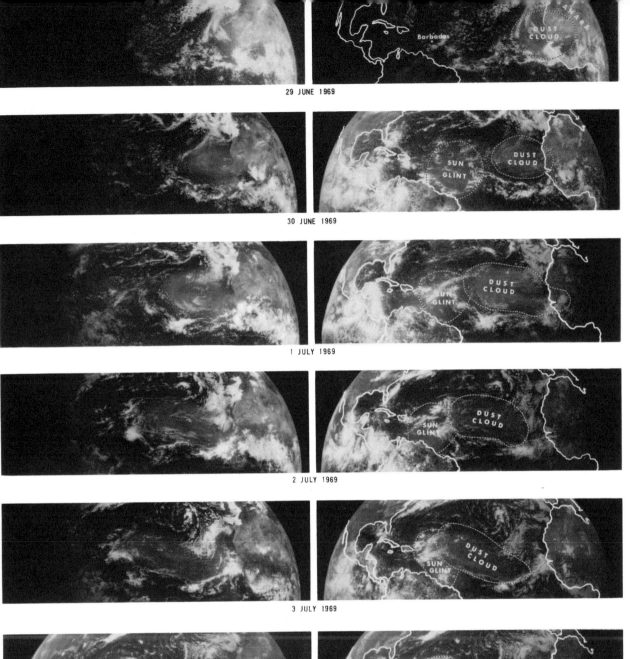

29 JUNE 1969

30 JUNE 1969

1 JULY 1969

2 JULY 1969

3 JULY 1969

4 JULY 1969

In six days in 1969 a dust cloud moved from the Sahara to the Caribbean as shown in this sequence of ATS 3 photographs. Such dust clouds can increase significantly the amount of solar radiation absorbed by the atmosphere, thus reducing the amount of solar energy that reaches the surface. The effects of dust clouds upon weather are not fully understood, but satellite reconnaissance should help to clarify the interrelationships involved. (NASA)

This stratocumulus formation, seen over the Pacific from Apollo 6, was of special interest to meteorologists because of the sharp line from top to bottom center. To its left, clouds are tightly clustered; to its right, they are more scattered. The line may mark the edge of a cold current or an area of upwelling over warm water. (NASA)

"This beautiful vortex is typical of the weather phenomena that can be seen from space," said astronaut Thomas P. Stafford of this photo, taken on his Gemini 6 flight. Island in upper left is Tenerife of the Canary group; eddies form when northeast winds, under temperature inversion layer, blow past the Canaries in presence of stratocumulus clouds. (NASA)

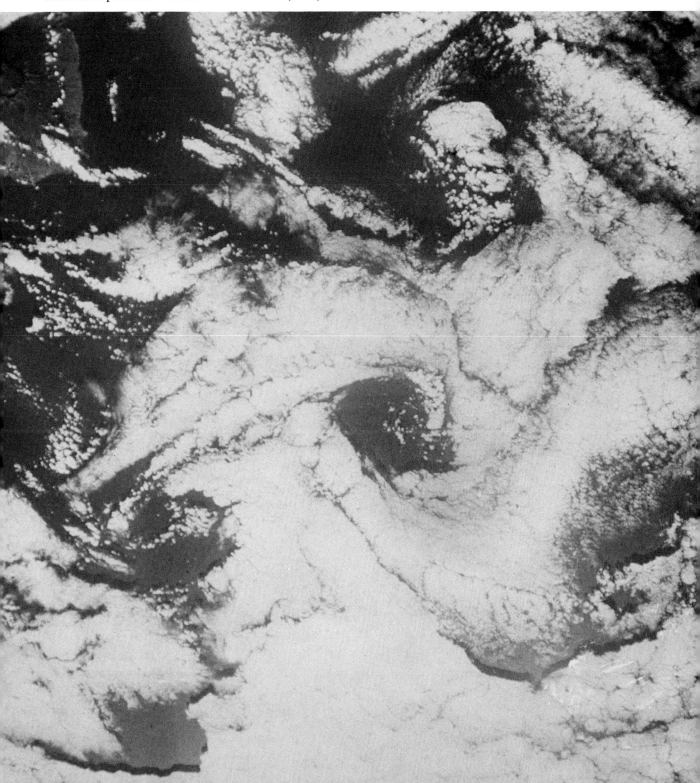

Jet stream cirrus clouds running across Nile Valley and Red Sea. These clouds occur near the strong core of upper westerly winds flowing between 35,000 and 40,000 feet. (NASA)

166

Although recent years have produced spectacular advances in meteorological and upper atmospheric research and applications, much remains to be accomplished before a truly global operational atmospheric observation system is achieved. A major step forward is the development of NOAA's geostationary operational environmental satellite, or GOES. With its 16-inch aperture telescope, this twenty-four-hour orbit spacecraft is capable of day and night visible and infrared scanning of weather phenomena, including tropical storms and hurricanes and midlatitude storms. Moreover, it can interrogate remotely located environmental observing platforms such as ocean buoys, river gages, ships, balloons, and possibly aircraft. Among other tasks, GOES is designed to provide communications links for the Pacific Tsunami Warning System, relaying data from networks of seismometers and tidal gages to NOAA's Honolulu Observatory and for regional Satellite Field Services Stations in Suitland, Maryland, Miami, Kansas City, San Francisco, and Hawaii (to provide short-term forecasts and warnings of storms, gales, hurricanes and tornados). For complete global coverage, four GOES satellites are necessary, complementing Itos–NOAA satellites following lower-altitude polar orbits.

"All sorts of things and weather / Must be taken in together," wrote Ralph Waldo Emerson. For a long time to come, weather will have to be taken in together with many other natural phenomena, but with the aid of the U.S. National Operational Environmental Satellite system and satellites flown by other countries, man is no longer a blind and helpless bystander. Not so many years ago, a hurricane such as Camille would have roared in from the Gulf of Mexico with little or no warning. But today, these incredibly energetic and destructive disturbances are tracked hour by hour from space as they approach the inhabited coastline and head inland. We may not yet know all the details of how the atmosphere works, but we can at least predict the weather with increasing accuracy—and some day may even be able to modify it to our tastes and profit.

One of the great lessons of history—and it applies with equal force to the use of satellites in all their many roles in geography, geology, hydrology, agriculture, and oceanography, as well as meteorology—is that man is almost invariably pessimistic when attempting to predict the economic impact of important technological innovations. And the psychological impact, so very significant in the case of the new vantage point man has gained from orbit, has been completely ignored more often than not.

Even to those most intimately involved in the space program, the change in the point of view can produce a jarring effect. This was evident from an incident that took place during the Apollo 8 mission in December 1968. Reporting on a conversation with the astronauts—the first men to go into orbit around the Moon—the mission control commentator said:

Certainly the quote that stopped us all, more than anything else came from [mission Commander Frank] Borman. I'm sure it was by accident, but at one point he, in trying [to execute a sequence] said "as soon as we find the Earth, we'll do it," and that brought a loud clap of laughter here.

167

Conventional notions of time, velocity, size, distance, and orientation suddenly seemed to change. As the three astronauts receded from Earth, so the Earth also receded in size. Yet planet Earth remained as marvelous as ever. Another astronaut on a later mission, Thomas P. Stafford on Apollo 10, in May 1969, conveyed the feeling:

Earth looked like a boy's blue glass marble—mottled black, gray, blue, and white against the blackness of the sky. It looks beautiful going away, and it'll look even better coming back.

Crescent Earth. Apollo 4, from about 11,000 miles out. (NASA)

# Appendix

## TYPICAL APPLICATIONS OF REMOTE SENSORS*

| Sensors | Agriculture and Forestry | Hydrology | Geography and Geodesy | Geology | Oceanography |
|---|---|---|---|---|---|
| Photography— visual | Crop and soil identification Identification of plant vigor and disease | Identification of drainage patterns | Urban-rural land use, transportation routes and facilities | Identification of surface structure | Identification of sea state, beach erosion, offshore depth and turbidity |
| Photography— multispectral | Crop and soil identification Identification of plant vigor and disease | Soil moisture content | Terrain and vegetation characteristics | Identification of surface features | Sea color as correlated with productivity |
| Infrared imagery and spectroscopy | Terrain composition Plant vigor and disease condition | Detection of areas cooled by evaporation | Surface energy budgets, near shore currents and land use | Mapping thermal anomalies Mineral identification | Mapping of ocean currents Sea-ice investigations |
| Radar imagery | Soil characteristics | Measurement of soil Moisture content Identification of runoff slopes | Land/ice mapping cartographic and geodetic mapping | Surface roughness Tectonic mapping | Sea state Ice flow and ice penetration Tsunami warning |
| Radio frequency, reflectivity | Soil characteristics | Moisture content of soils | Land/ice mapping and thickness Penetration of vegetation cover | Sub-surface layering Mineral identification | Sea ice thickness and mapping Sea state |
| Passive microwave radiometry and imagery | Brightness temperature Map of terrain | Snow and ice surveys | Snow and ice measurements | Dielectric constant Measurement indicative of subsurface layering | |
| Absorption spectroscopy | | | | Detection of mineral deposits Trace metals, and oil fields | Detection of concentrations of surface marine flora |

* Adopted from NASA Headquarters Manned Space Sciences and Applications, Washington, and NASA Manned Spacecraft Center, Houston, Texas.

## EARLY SATELLITE METEOROLOGICAL EXPERIMENTS

| Satellite | Date of Launch | Initial Orbit, perigee-apogee, miles | | Experiment |
|---|---|---|---|---|
| Vanguard 2 | 17 February 1959 | 347 | 2,064 | Two infrared-sensitive photocells designed to receive light collected by a pair of 3-in. parabolic mirrors, or scanners. Solar radiation reflections from cloud formations, as well as ocean and land surfaces, were translated into electrical impulses and stored on magnetic tape for readout upon interrogation. The spin of the satellite served as the scanning technique. |
| Explorer 6 | 7 August 1959 | 156 | 26,357 | TV-type optical scanner to relay crude pictures of cloud cover. |
| Explorer 7 | 13 October 1959 | 347 | 702 | Radiometer to investigate solar radiation balance (heat budget) of the Earth. Six sensing elements employed: two black hemispheres measured energy over all wavelengths; and two Tabor sensors (with special coatings to make them absorption sensitive to short wavelengths) measured short wavelengths. |

## U.S. UNMANNED APPLICATIONS SATELLITES
### 1960–1974

| Spacecraft | Launch Date | Launch Vehicle | Remarks |
|---|---|---|---|
| Tiros 1 | 1 April 1960 | Thor Able | First weather satellite providing cloud cover photography. Operated 79 days; produced 19,389 pictures. |
| Tiros 2 | 23 November 1960 | Thor Delta | 25,574 pictures taken; full operations for 69 days, sporadic operations until 12 July 1961. |
| Tiros 3 | 12 July 1961 | Thor Delta | 24,000 pictures produced during 108 days, and 70 storm advisories given. |
| Tiros 4 | 8 February 1962 | Thor Delta | 23,370 pictures taken during 125 days, and 102 storm advisories given. |
| Tiros 5 | 19 June 1972 | Thor Delta | 48,547 pictures, 399 storm advisories during 320 operational days. |
| Tiros 6 | 18 September 1962 | Thor Delta | 59,830 pictures, 361 storm advisories during 388 operational days. |

| Spacecraft | Launch Date | Launch Vehicle | Remarks |
|---|---|---|---|
| Tiros 7 | 19 June 1963 | Thor Delta | 111,047 pictures, 596 storm advisories. Ended operations on 3 February 1966. |
| Tiros 8 | 21 December 1963 | Thor Delta | 88,662 pictures, 424 storm advisories. Abandoned on 1 July 1967. First satellite to transmit continuously local cloud conditions to APT (automatic picture transmission) stations. |
| Nimbus 1 | 28 August 1964 | Thor Agena B | Carried advanced vidicon system, high resolution infrared radiometers for night pictures, and APT-operated for 2 months; tracked many hurricanes and typhoons. |
| Tiros 9 | 22 January 1965 | Thor Delta | First satellite in Sun-synchronous orbit. Wheel configuration. 76,604 pictures and 562 storm advisories to 15 February 1967. |
| Tiros 10 | 2 July 1965 | Thor Delta | 59,119 pictures, 307 storm advisories during 732 operational days. Abandoned 3 July 1967. |
| ESSA 1 | 3 February 1966 | Thor Delta | First operational weather satellite; carried two wide-angle TV cameras 101,938 pictures, 631 storm advisories during 162 days. |
| ESSA 2 | 28 February 1966 | Thor Delta | Complemented ESSA 1 with two wide-angle APT cameras; ended operations on 16 October 1970. Pictures via local readout. |
| Nimbus 2 | 15 May 1966 | Thor Agena B | Continued development of Nimbus satellite system. In first two months alone, took more than 150,000 pictures and received more than 23,000 commands from ground controllers. Made available for first time instantaneous round-the-clock, worldwide weather pictures. Operated for almost 33 months, terminating on orbit No. 13,029. |
| ESSA 3 | 2 October 1966 | Thor Delta | Odd numbered craft carried two advanced videcon camera systems. 91,272 pictures, 1,971 storm advisories over 241 operational days. |
| ATS 1 | 6 December 1966 | Atlas Agena D | Continuous black-and-white cloud-cover pictures from synchronous orbit. |
| ESSA 4 | 26 January 1967 | Thor Delta | Even-numbered spacecraft carried two APT camera systems. Local readout, 116 operational days. |
| ESSA 5 | 20 April 1967 | Thor Delta | 1,037 operational days; 79, 301 pictures, 788 storm advisories. |
| ATS 3 | 5 November 1967 | Atlas Agena D | Continuous color cloud cover pictures from synchronous orbit. |
| ESSA 6 | 10 November 1967 | Thor Delta | 725 operational days; local readout. |
| ESSA 7 | 16 August 1968 | Thor Delta | 338 operational days; 79,470 pictures, and 944 storm advisories. |

| Spacecraft | Launch Date | Launch Vehicle | Remarks |
|---|---|---|---|
| ESSA 8 | 15 December 1968 | Thor Delta | 777 operational days; local readout. |
| ESSA 9 | 26 February 1969 | Thor Delta | 704 operational days, 67,362 pictures through 31 January 1971 after which records of such pictures no longer maintained. |
| Nimbus 3 | 14 April 1969 | Thor Agena D | First vertical temperature and water vapor profile on global basis of atmosphere down to the Earth's surface. Conducted many oceanographic experiments, and even tracked an elk in Wyoming. |
| ITOS 1 (Tiros M) | 23 January 1970 | Thor Delta | First launch of second generation meteorological satellite. 510 operational days. |
| Nimbus 4 | 8 April 1970 | Thorad Agena D | Carried for first time new instruments to monitor distribution of atmospheric ozone, of water vapor, and temperatures from cloud top level to about 40 miles high as well as improved instruments used in earlier Nimbus. |
| NOAA 1 (ITOS A) | 11 December 1970 | Thor Delta | Same payload as ITOS 1; used to augment U.S. weather satellite capability. Operated for 25 days. |
| ERTS 1 | 23 July 1972 | Thor Delta | Returned repetitive, multi-spectral images of use to the disciplines of oceanography, agriculture, forestry, geography, geology, and hydrology. In first two years 100,000 pictures received, used by investigators in 49 nations. |
| NOAA 2 (ITOS D) | 15 October 1972 | Thor Delta | Same as NOAA 1, covering global atmosphere and oceans twice each day. First operational weather satellite to rely entirely on scanning radiometers for imagery. |
| Nimbus 5 | 11 December 1972 | Thor Delta | First atmospheric vertical temperature profile measurements through clouds. Successfully measured ocean rainfall on global scale, and studied meandering of the Gulf Stream. Also on board: an Earth resources experiment designed to measure the temperature of the Earth's crust twice a day. |
| NOAA 3 | 6 November 1973 | Thor Delta | Provided visible and infrared images of cloud cover, snow and ice, and the ocean surface. Also monitored infrared radiation at 6 atmospheric levels, at Earth's surface, or, in overcast areas, of cloud tops. |
| SMS 1 | 17 May 1973 | Thor Delta | From some 22,500 miles above coastal Brazil the first Synchronous Meteorological Satellite provided day and night pictures of the western hemisphere, monitors solar flare activity, and received and relayed environmental data from some 10,000 manned and unmanned data collection platforms. |

## SUMMARY OF THE SKYLAB FLIGHT PROGRAM

|  | *SL–1* | *SL–2* | *SL–3* | *SL–4* | *Totals* |
|---|---|---|---|---|---|
| Launch date | 14 May 1973 | 25 May 1973 | 28 July 1973 | 16 November 1973 | |
| Carrier vehicle | Saturn 5 (AS–513) | Saturn 1B (AS–206) | Saturn 1B (AS–207) | Saturn 1B (AS–208) | |
| Recovery date | expected to decay into atmosphere around 1982 | 22 June 1973 | 25 September 1973 | 8 February 1974 | |
| Duration of mission, days-hours-minutes | | 28–0–49 | 59–11–9 | 84–1–16 | 171–13–14 |
| Distance traveled, manned, in miles | | 11.5 | 24.5 | 34.5 | 70.5 |
| Number of revolutions, manned | | 404 | 858 | 1,214 | 2,476 |
| Crew: | | | | | |
| Commander | | Conrad | Bean | Carr | |
| Pilot | | Weitz | Lousma | Pogue | |
| Scientist pilot | | Kerwin | Garriott | Gibson | |

## EARLY EARTH-LOOK PHOTOGRAPHY FROM MANNED SPACECRAFT[*]

| *Launch vehicle* | *Spacecraft* | *Date launched* | *Photography* |
|---|---|---|---|
| Redstone | Mercury MR3 | 5 May 1961 | 200 frames of 70mm color photography using fixed Mauer camera; scenes of clouds and water. |
| Redstone | Mercury MR4 | 21 July 1961 | Although photographs were taken during the flight, the spacecraft was lost at sea following splashdown and camera and film were lost. |
| Atlas | Mercury MA6 | 20 February 1962 | 38 frames of 35mm color film showing clouds, water and terrain. |
| Atlas | Mercury MA7 | 24 May 1962 | 200 frames of 35mm color film showing clouds, water, and terrain. |
| Atlas | Mercury MA8 | 3 October 1962 | 14 frames of 70mm hand-held Hasselblad camera showing clouds, water, and terrain. |
| Atlas | Mercury MA9 | 15 May 1963 | 31 frames of 70mm hand-held Hasselblad camera showing clouds, water, and terrain. |

[*] Based on data supplied by Olav Smistad, Manager, Operations Planning and Requirements Office, Earth Resources Program, NASA—Lyndon B. Johnson Space Center, Houston.

## VOLUME OF EARTH-LOOK PHOTOGRAPHY DURING
## GEMINI, APOLLO, AND SKYLAB MISSIONS

| *Spacecraft* | *Number of frames* |
| --- | --- |
| Gemini 3 | 25 |
| Gemini 4 | 205 |
| Gemini 5 | 235 |
| Gemini 6 | 160 |
| Gemini 7 | 410 |
| Gemini 8 | 18 |
| Gemini 9 | 322 |
| Gemini 10 | 325 |
| Gemini 11 | 220 |
| Gemini 12 | 371 |
| Apollo 4 (unmanned capsule test) | 669 |
| Apollo 6 (unmanned capsule test) | 372 |
| Apollo 7 | 443 |
| Apollo 8 | 51 |
| Apollo 9 | 1,157 |
| Apollo 10 | 93 |
| Apollo 11 | 141 |
| Apollo 12 | 100 |
| Skylab 1 | 9,846 |
| Skylab 2 | 16,800 |
| Skylab 3 | 19,500 |

# Bibliography

Some five hundred basic sources of information were consulted during the preparation of this book, ranging from massive multivolume proceedings containing over one thousand pages to short articles bearing such titles as "Concealed Structures in Arctic Alaska Identified on ERTS-1 Imagery" (in *Oil and Gas Journal*, May 28, 1973) to single-page information releases. It not being feasible to publish herein anything approaching a complete reference list, the author elected rather to include only representative sources, particularly those of a compilatory and review nature.

Annual Earth Resource Program Reviews. Houston, 1968, 1969, 1970, and 1972: NASA Manned Spacecraft (now Johnson Space) Center. (All are four-volume works.)

Center for Short-Lived Phenomena Annual Reports. Cambridge, Mass.: Smithsonian Institution.

"Early Results from Skylab 1." In *Astronautics and Aeronautics* 11, No. 9 (September 1973), 22

*Earth Photographs from Gemini III, IV, and V; and Earth Photographs from Gemini VI Through XII.* Washington, D.C., 1967 and 1968: NASA Reports SP-129 and SP-171.

*Earth Resources: Cooperative Research in Remote Sensing for Earth Surveys—Agreement Between United States and Brazil.* Washington, D.C., 1968. Dept. of State Publication 6569. Also, between the U.S. and Mexico, publication 6613.

Earth Resources Program Synopsis of Activity. Houston, 1970: NASA—Manned Spacecraft (now Johnson Space) Center.

Earth Resources Survey Systems. Washington, D.C., 1972: NASA Report SP-283, (Proceedings of international workshop sponsored by NASA; U.S. Departments of Agriculture, Interior, State and Defense; National Oceanic and Atmospheric Administration, Naval Oceanographic Office, and Agency for International Development.)

Earth Resources Satellite System. Washington, D.C., 1969: U.S. Government Printing Office. (Report for the Subcommittee on NASA Oversight of the Committee of Science and Astronautics, U.S. House of Representatives; Serial W.)

Ecological Surveys from Space. Washington, D.C., 1970: NASA Report SP-230.

Exploring Space with a Camera. Washington, D.C., 1968: NASA Report SP-138.

"Infrared Magic," *Agricultural Research* 18, No. 1 (July 1969), 8.

Man's Geophysical Environment: Its Study from Space. Washington, D.C., 1969: U.S. Government Printing Office. (Contains sections on space disturbances, aeronomy, meteorology, hydrology, oceanography, and hydrography, geodesy, and systems considerations—including the role of man.)

NASA Authorizations: Hearings before the Committee on Aeronautical and Space Sciences of the United States Senate (annual). Volumes published between the late 1960s and 1974 provide excellent summaries of work involved in the study of the planet Earth from orbit. Washington, D.C.: U.S. Government Printing Office.

NASA Authorizations: Hearings before the Subcommittee on Space Science and Applications of the Committee on Science and Astronautics of the House of Representatives published during recent years provide excellent summaries of work involved in the study of the planet Earth from orbit. Washington, D.C.: U.S. Government Printing Office.

Peaceful Uses of Earth-Observation Spacecraft. Washington, D.C., 1966: NASA Reports CR–586, –587 and –588 (prepared by University of Michigan at Ann Arbor; three volumes).

Proceedings, Photo-optical Instrumentation Engineers, 14th Annual Technical Symposium and Exhiborama. Redondo Beach, California, 1969: Society of Photo-optical Instrumentation Engineers. (Contains sections on imaging sensors, infrared systems, remote sensing, etc.)

Proceedings of Symposia on Remote Sensing of Environment. Ann Arbor, Michigan: Willow Run Laboratories and Institute of Science and Technology of the University of Michigan. (Through 1974 nine of these multidisciplinary proceedings were published.)

Reference Earth Orbital Research and Applications Investigations—Volume IV, Earth Observations. Washington, D.C., 1971: NASA Report NHB 7150.1.

Remote Sensing of Earth Resources. Washington, D.C., 1970: NASA Report SP-7036 (literature survey with indexes).

Significant Achievements in Satellite Meteorology, 1958–1964, Washington, D.C., 1966: NASA Report SP-96.

Skylab Experiments–Volume 2, Remote Sensing of Earth Resources. Washington, D.C., 1973: U.S. Government Printing Office.

Space Environmental Vantage Point. Washington, D.C., 1971: U.S. Government Printing Office. (Prepared by the National Oceanic and Atmospheric Administration.)

A Survey of Space Applications. Washington, D.C., 1967: NASA Report SP-142.

Symposium on Significant Results Obtained from the Earth Resources Technology Satellite-1. Volume I, Sections A and B, Technical Presentations; Volume II, Summary of Results; and Volume III, Discipline Summary Reports. Volume I; Washington, 1973: NASA; Volumes II and III: Greenbelt, Md., 1973: NASA-Goddard Space Flight Center.

Useful Applications of Earth-Oriented Satellites. Washington, D.C., 1969: National Academy of Sciences. Consists of "Summaries of Panel Reports," "Report of the Central Review Committee," and individual panel reports as follows: (1) "Forestry-Agriculture-Geography," (2) "Geology," (3) "Hydrology," (4) "Meteorology," (5) "Oceanography," (6) "Sensors and Data Systems," (7) "Point-to-point Communication," (10) "Broadcasting," (11) "Navigation and Traffic Control," and (12) "Economic Analysis."

Alfred, John C., "New Concept: Sensing Oceans from Space." In *Ocean Industry* 9 No. 6 (June 1974), 59.


Badgley, Peter.; Miloy, Leatha; and Childs, Leo F. *Oceans from Space.* Houston, Texas, 1970: Gulf Publishing Co.

Badgley, Peter C., ed. *Scientific Experiments for Manned Orbital Flight—Science and Technology Series*-Volume 4. North Hollywood, Calif., 1965: Western Periodicals Co.

Barnes, James C. and Bowley, Clinton J., "Mapping Sea Ice from the Earth Resources Technology Satellite." In *Arctic Bulletin* 1, No. 1 (Summer 1973), 6.

Belew, Leland F. and Stuhlinger, Ernst. *Skylab: A Guide-book.* Washington, D.C., 1973: U.S.

Government Printing Office. (See Section V-2, "Earth Resources Experiment Program.")

Bodechiel, Johann and Gierloff-Emden, Hans Günter, *The Earth from Space.* New York: Arco Publishing Co., 1974.

Bylinsky, Gene. "From a High-Flying Technology, A Fresh View of Earth," In *Fortune* 78 No. 7 (1 June 1968).

Colwell, Robert N. "Remote Sensing of Natural Resources," *Scientific American* 218. No. 1 (January 1968), 54.

Corliss, William R. *Satellites at Work.* Washington, D.C., 1971: NASA Publication EP-84.

Dornbach, John E. Analysis of Apollo AS-501 Mission Earth Photography. Houston, 1968: NASA Report TM X-58015.

Enzmann, Robert Duncan, ed. *Use of Space Systems for Planetary Geology and Geophysics— Science and Technology Series.* Volume 17. Tarzana, Calif., 1968: American Astronautical Society Publications Office.

Ewing, Clifford C. *Oceanography from Space— Proceedings of Conference on the Feasibility of Conducting Oceanographic Explorations from Aircraft, Manned Orbital and Lunar Laboratories.* Woods Hole, Massachusetts, 1965: Report Woods Hole Oceanographic Institution Feb. 65–10.

Fischer, William A. "Remote Sensing Research in the United States." In *Proceedings of the XL International Congress for Photogrammetry.* Lausanne, 1968 (available through U.S. Geological Survey, Washington).

Fitzgerald, Ken. *The Space-Age Photographic Atlas,* New York, 1970, Crown.

Fletcher, James C. ERTS-1: Toward Global Monitoring. Same journal, p. 32. Also, Mercanti, Enrico P., ERTS-1: Teaching Us a New Way to See. Same journal, p. 36.

Gerlach, Arch C. "The Geographical Applications Program of the U.S. Geological Survey," *Photogrammetric Engineering* 35, No. 1 (January 1969), 58.

Gilmer, J. Ray; Mayo, A. M.; and Peavey, R. C.

eds. *Commercial Utilization of Space—Advances in the Astronautical Sciences* Volume 23. Tarzana, Calif., 1968: AAS Publications Office.

Graham, L. C. "Earth Resources Determination with Terrain Imaging Radars." In Burly, R., ed., *Resources Roundup.* Phoenix, Arizona, 1969: Institute of Electrical Engineers.

Haggerty, James J. "The Giant Harvest from Space —Today and Tomorrow." In *Air Force and Space Digest* 53, No. 2 (February 1970), 30.

Holliday, C. T. "The Earth as Seen from Outside the Atmosphere." In *Earth as a Planet* (ed. by G. P. Kuiper). Chicago, 1954: University of Chicago Press.

Jaffe, Leonard. "NASA Earth Resources Satellite Research and Development Program." In Napolitano, L. G., ed., *Astronautical Research*, p. 785. Amsterdam, 1971: North-Holland.

Johnson, Arthur W. "Weather Satellites." In *Scientific American* 220, No. 1 (January 1969).

Kaltenbach, John L. Apollo 9 Multispectral Photographic Information. Washington, D.C., 1970: NASA Report TM X-1957.

———. Science Screening Report of the Apollo 7 Mission 70-Millimeter Photography and NASA Earth Resources Aircraft Mission 981 Photography. Houston, 1969: NASA Report TM X-58029.

———. Science Report on the 70-millimeter Earth Photography of the Apollo 6 Mission. Houston, 1969: NASA Report S-217.

Katz, Amron H. "Observation Satellites: Problems and Prospects." In *Astronautics* 5, No. 4 (April 1960), 26; 5 No. 6 (June 1960), 26; 5 No. 7 (July 1960), 28; 5 No. 8 (August 1960), 30; 5 No. 9 (September 1960), 32; and 5 No. 10 (October 1960), 36.

Kavanau, L. L., ed. *Practical Space Applications —Advances in the Astronautical Sciences*-Volume 21. Sun Valley, Calif., 1967; Scholarly Publications, Inc.

Lowman, Paul D. Jr., "Photography From Space," In *Science Journal* 1, No. 3 (May 1965), 52.

———. A Review of Photography of the Earth from Sounding Rockets and Satellites. Washington, D.C., 1964: NASA Report TN D-1868.

———. *Space Panorama*, Zürich, 1968: Weltflugbild.

———. "Space Photography—A Review." In *Photogrammetric Engineering* 31, No. 1 (January 1965), 76.

Ludwick, E. E., Jr. "Space Oceanography—Applications and Benefits." In Volume 2, *Proceedings of the Sixth Space Congress Canaveral Council of Technical Societies—Space, Technology and Society*. Cape Canaveral, Florida, 1969: Canaveral Council of Technical Societies.

Ludwig, E. B.; Bartle, R.; and Griggs, M. Study of Air Pollutant Detection by Remote Sensors, Washington, D.C., 1969: NASA Report CR 1380.

Mannella, Gene G., "Aerospace Sensor Systems." In *Astronautics and Aeronautics* 6, No. 12 (December 1968), 27.

Maughan, Paul M., "Remote-Sensor Applications in Fishery Research," *Marine Technology Society Journal* 3, No. 2 (March 1969), 11.

Mekel, J. F. M., *Geology from the Air*. Delft, Netherlands, 1969: Uitgevezij Waltman.

Mueller, George E., "Earthly Dividends from Space." In *Spaceflight* 11, No. 12 (December 1969), 418.

Narin, Francis, ed. *Post Apollo Space Exploration—Advances in the Astronautical Sciences-Volume 20*. Tarzana, Calif., 1966: American Astronautical Society Publications Office.

Newell, H. E., Current Programme and Considerations of the Future Earth Resources Survey," *Spaceflight*, 10, No. 8 (Aug. 1968), 281.

Nunnally, Nelson R., Bibliography of Remote Sensing Applications for Planning and Administrative Studies. Johnson City, Tenn., December 1971: East Tennessee State University, and Washington, D.C., U.S. Department of Interior-Geological Survey, Interagency Report USGS 234. (Published at the same time by the Geological Survey: "Preliminary EROS Program Bibliography.")

Olson, Boyd E., Remote Sensing in Oceanography. Washington, D.C., 1968: Naval Oceanographic Office Report IR 68-18.

Ordway, Frederick I. III "Think About This!—Geology has Great Potential in the Space Age." In *Geotimes* 2 No. 10 (April 1958), 12.

Ordway, Frederick I., III; Adams, Carsbie C.; and Sharpe, Mitchell R. *Dividends from Space*. New York, 1971: Thos. Y. Crowell.

Owtrow, Harvey and Weinstein, Oscar. "A Review of a Decade of Space Camera Systems Development for Meteorology." In *Proceedings, Society of Photo-Optical Instrumentation Engineers* 13th Annual Technical Symposium-Vol 1. Redondo Beach, Calif., 1969: Society of Photo-Optical Instrumentation Engineers.

Pecora, William T. "Surveying the Earth's Resources from Space." In *Surveying and Mapping* 27, No. 4 (April 1967), 639.

Poluquet, J. *Earth Sciences in the Age of the Satellite*. Dortrecht-Holland, 1974: D. Reidel Publishing Co.

Pesce, Angelo, *Gemini Space Photographs Of Libya and Tibesti*. Tripoli, 1968: Petroleum Exploration Society of Libya.

Puttkamer, Jesco von, and McCullough, Thomas J., eds. Space for Mankind's Benefit. Washington, D.C., 1972: NASA Report SL-313 (Proceedings of a congress held in Huntsville, Alabama).

Pyle, Robert L., "Weather Satellite Capabilities: Present and Future." In *Weatherwise* 25 No. 5 (October 1972), 209.

Roberts, Walter Orr, "We're Doing Something about the Weather," In *National Geographic* 141, No. 4 (April 1972), 518.

Robinove, Charles J., "Perception via Satellite." In *Water Spectrum* 2, No. 1 (Spring 1970), 14.

Ruppe, Harry O., "Astronautics: An Outline of Utility." In Ordway, Frederick I., ed., *Advances in Space Science and Technology—*

Volume 10. New York, 1970: Academic Press.

Schneider, William C. and Green, William D. Jr., "The Skylab Experiment Program." In Ordway, Frederick I., ed., *Advances in Space Science and Technology*—Volume 11. New York, 1972: Academic Press. (See Section IV, "Earth Observations.")

Smith, John T., ed., *Manual of Color Aerial Photography*. New York, 1968: American Society of Photogrammetry.

Stehling, Kurt R., "Spotting Pollution from Space," *Space Aeronautics*, 53 No. 6 (June 1970), 47.

Stevenson, Robert E., A Real-Time Fisheries Satellite System. Galveston, Texas, 1969: Bureau of Commercial Fisheries Biological Laboratory Contribution No. 287.

Stoertz, George E., William R. Hemphill, and David A. Markle, "Airborne Fluorometer Applicable to Marine and Estuarine Studies." In *Marine Technology Society Journal* 3, No. 6 (November–December 1969), 11.

Vaeth, J. Gordon, "Establishing an Operational Weather Satellite System." In Ordway, Frederick I., ed., *Advances in Space Science and Technology*-Volume 7. New York, 1965: Academic Press.

Warnecke, Guenter; McMillin, Larry M.; and Allison, Lewis J. Ocean Current and Sea Surface Temperature Observations from Meteorological Satellites. Washington, D.C., 1969: NASA Report TN D-5142.

Widger, W. K., *Meteorological Satellites*, New York, 1966: Holt, Rinehart and Winston, 1966.

Williams, O. W., "Surveying the Earth by Satellite." In *Science Journal*, 3, No. 1, (January 1967), 68.

# Index

183

190